NATEF Standards Job Sheets

Manual Transmissions (A3)

Fourth Edition

Jack Erjavec

Ken Pickerill

CENGAGE
Learning·

Australia • Brazil • Japan • Korea • Mexico • Singapore • Spain • United Kingdom • United States

CENGAGE
Learning

NATEF Standards Job Sheets
Manual Transmissions (A3)
Fourth Edition
Jack Erjavec & Ken Pickerill

VP, General Manager, Skills and Planning:
Dawn Gerrain

Director, Development, Global Product
Management, Skills: Marah Bellegarde

Product Manager: Erin Brennan

Senior Product Development Manager:
Larry Main

Senior Content Developer: Meaghan Tomaso

Product Assistant: Maria Garguilo

Marketing Manager: Linda Kuper

Market Development Manager:
Jonathon Sheehan

Senior Production Director: Wendy Troeger

Production Manager: Mark Bernard

Content Project Management: S4Carlisle

Art Direction: S4Carlisle

Media Developer: Debbie Bordeaux

Cover image(s): © Snaprender/Dreamstime.com

© Shutterstock.com/gameanna

© i3alda/www.fotosearch.com

© IStockPhoto.com/tarras79

© IStockPhoto.com/tarras79

© Laurie Entringer

© IStockPhoto.com/Zakai

© stoyanh/www.fotosearch.com

For product information and technology assistance, contact us at
Cengage Learning Customer & Sales Support, 1-800-354-9706

For permission to use material from this text or product,
submit all requests online at **www.cengage.com/permissions.**
Further permissions questions can be e-mailed to
permissionrequest@cengage.com

Library of Congress Control Number: 2014931171

ISBN-13: 978-1-1116-4699-8
ISBN-10: 1-111-64699-6

Cengage Learning
200 First Stamford Place, 4th Floor
Stamford, CT 06902
USA

Cengage Learning is a leading provider of customized learning solutions with office locations around the globe, including Singapore, the United Kingdom, Australia, Mexico, Brazil, and Japan. Locate your local office at:
www.cengage.com/global

Cengage Learning products are represented in Canada by Nelson Education, Ltd.

To learn more about Cengage Learning, visit **www.cengage.com**
Purchase any of our products at your local college store or at our preferred online store **www.cengagebrain.com**

Notice to the Reader

Publisher does not warrant or guarantee any of the products described herein or perform any independent analysis in connection with any of the product information contained herein. Publisher does not assume, and expressly disclaims, any obligation to obtain and include information other than that provided to it by the manufacturer. The reader is expressly warned to consider and adopt all safety precautions that might be indicated by the activities described herein and to avoid all potential hazards. By following the instructions contained herein, the reader willingly assumes all risks in connection with such instructions. The publisher makes no representations or warranties of any kind, including but not limited to, the warranties of fitness for particular purpose or merchantability, nor are any such representations implied with respect to the material set forth herein, and the publisher takes no responsibility with respect to such material. The publisher shall not be liable for any special, consequential, or exemplary damages resulting, in whole or part, from the readers' use of, or reliance upon, this material.

Printed in the United States of America
1 2 3 4 5 6 7 18 17 16 15 14

CONTENTS

NATEF Task List for Manual Drive Train and Axles   **v**

    Required Supplemental Tasks (RST) List   v
    Maintenance and Light Repair (MLR) Task List   vi
    Automobile Service Technology (AST) Task List   vi
    Master Automobile Service Technology (MAST) Task List   viii

Definition of Terms Used in the Task List   **x**

Cross-Reference Guides   **xiii**

Preface   **xvii**

Manual Transmissions   **1**

    Basic Manual Transmission Theory   1
    Safety   8
    Hazardous Materials and Wastes   12
    Manual Transmission Tools and Equipment   14

NATEF RST Job Sheets   **21**

    Job Sheet 1   Shop Safety Survey   21
    Job Sheet 2   Working in a Safe Shop Environment   25
    Job Sheet 3   Fire Extinguisher Care and Use   27
    Job Sheet 4   Working Safely Around Air Bags   29
    Job Sheet 5   High-Voltage Hazards in Today's Vehicles   31
    Job Sheet 6   Hybrid High-Voltage and Brake System Pressure Hazards   33
    Job Sheet 7   Material Data Safety Sheet Usage   37
    Job Sheet 8   Measuring Tools and Equipment Use   39
    Job Sheet 9   Preparing the Vehicle for Service and Customer   41

Manual Drive Train and Axles NATEF MLR/AST/MAST Job Sheets   **45**

    Job Sheet 10   Identifying Problems and Concerns   45
    Job Sheet 11   Gathering Vehicle Information   49
    Job Sheet 12   Drive Train Fluid Service   53
    Job Sheet 13   Troubleshoot a Clutch Assembly   57
    Job Sheet 14   Clutch Linkage Inspection and Service   61
    Job Sheet 15   Clutch Inspection and Service   65
    Job Sheet 16   Hydraulic Clutch Service   71
    Job Sheet 17   Inspect and Adjust Shift Linkage   75
    Job Sheet 18   Electronic Controlled Transmissions   79
    Job Sheet 19   Diagnosing Noise Problems   81
    Job Sheet 20   Road-Test a Vehicle for Transmission Problems   87
    Job Sheet 21   Disassemble and Reassemble a Typical Transaxle   89
    Job Sheet 22   Drive Axle Inspection and Diagnosis   93

Job Sheet 23   Diagnosing RWD Noise and Vibration Concerns   97
Job Sheet 24   Servicing FWD Wheel Bearings   105
Job Sheet 25   Servicing Universal Joints and CV Joints   109
Job Sheet 26   Drive Axle Leak Diagnosis   117
Job Sheet 27   Differential Housing Service   121
Job Sheet 28   Diagnosing Differential Noise and Vibration Concerns   125
Job Sheet 29   Companion Flange and Pinion Seal Service   129
Job Sheet 30   Measure and Adjust Pinion Depth, Bearing Preload, and Backlash   133
Job Sheet 31   Differential Case Service   137
Job Sheet 32   Limited-Slip Differential Diagnostics   139
Job Sheet 33   Axle Shaft and Bearing Service   141
Job Sheet 34   Servicing Transfer Case Shift Controls   147
Job Sheet 35   Inspecting Front-Wheel Bearings and Locking Hubs   151
Job Sheet 36   Check Fluid in a Transfer Case   153
Job Sheet 37   Tire Size Changes   155
Job Sheet 38   Road-Test a Transfer Case   159
Job Sheet 39   Servicing 4WD Electrical Systems   161
Job Sheet 40   Disassemble and Reassemble a Transfer Case   167

NATEF TASK LIST FOR MANUAL DRIVE TRAIN AND AXLES

Required Supplemental Tasks (RST) List

Shop and Personal Safety

1. Identify general shop safety rules and procedures.

2. Utilize safe procedures for handling of tools and equipment.

3. Identify and use proper placement of floor jacks and jack stands.

4. Identify and use proper procedures for safe lift operation.

5. Utilize proper ventilation procedures for working within the lab/shop area.

6. Identify marked safety areas.

7. Identify the location and the types of fire extinguishers and other fire safety equipment; demonstrate knowledge of the procedures for using fire extinguishers and other fire safety equipment.

8. Identify the location and use of eyewash stations.

9. Identify the location of the posted evacuation routes.

10. Comply with the required use of safety glasses, ear protection, gloves, and shoes during lab/shop activities.

11. Identify and wear appropriate clothing for lab/shop activities.

12. Secure hair and jewelry for lab/shop activities.

13. Demonstrate awareness of the safety aspects of supplemental restraint systems (SRS), electronic brake control systems, and hybrid vehicle high-voltage circuits.

14. Demonstrate awareness of the safety aspects of high-voltage circuits (such as high intensity discharge (HID) lamps, ignition systems, injection systems, etc.).

15. Locate and demonstrate knowledge of material safety data sheets (MSDS).

Tools and Equipment

1. Identify tools and their usage in automotive applications.

2. Identify standard and metric designation.

3. Demonstrate safe handling and use of appropriate tools.

4. Demonstrate proper cleaning, storage, and maintenance of tools and equipment.

5. Demonstrate proper use of precision measuring tools (i.e., micrometer, dial-indicator, dial-caliper).

Preparing Vehicle for Service

1. Identify information needed and the service requested on a repair order.

2. Identify purpose and demonstrate proper use of fender covers, mats.

3. Demonstrate use of the three C's (concern, cause, and correction).

4. Review vehicle service history.

5. Complete work order to include customer information, vehicle identifying information, customer concern, related service history, cause, and correction.

Preparing Vehicle for Customer

1. Ensure vehicle is prepared to return to customer per school/company policy (floor mats, steering wheel cover, etc.).

Maintenance and Light Repair (MLR) Task List

A. General
A.1. Research applicable vehicle and service information, fluid type, vehicle service history, service precautions, and technical service bulletins. Priority Rating 1
A.2. Drain and refill manual transmission/transaxle and final drive unit. Priority Rating 1
A.4. Check fluid condition; check for leaks. Priority Rating 2

B. Clutch
B.1. Check and adjust clutch master cylinder fluid level. Priority Rating 1
B.2. Check for system leaks. Priority Rating 1

C. Transmission/Transaxle
C.1. Describe the operational characteristics of an electronically controlled manual transmission/transaxle. Priority Rating 3

D. Drive Shaft and Half Shaft, Universal and Constant-Velocity (CV) Joints
D.1. Inspect, remove, and replace front wheel drive (FWD) bearings, hubs, and seals. Priority Rating 2
D.2. Inspect, service, and replace shafts, yokes, boots, and universal/CV joints. Priority Rating 2

E. Differential Case Assembly
E.1. Clean and inspect differential housing; check for leaks; inspect housing vent. Priority Rating 2
E.2. Check and adjust differential housing fluid level. Priority Rating 1
E.3. Drain and refill differential housing. Priority Rating 1
    E.1.1. *Drive Axles*
        1. Inspect and replace drive axle wheel studs. Priority Rating 2

F. Four-Wheel Drive/All-Wheel Drive
F.1. Inspect front-wheel bearings and locking hubs. Priority Rating 3
F.2. Check for leaks at drive assembly seals; check vents; check lube level. Priority Rating 2

Automobile Service Technology (AST) Task List

A. General: Drive Train Diagnosis
A.1. Identify and interpret drive train concerns; determine necessary action. Priority Rating 1
A.2. Research applicable vehicle and service information, fluid type, vehicle service history, service precautions, and technical service bulletins. Priority Rating 1

A.3.   Check fluid condition; check for leaks; determine necessary action.   Priority Rating 1

A.4.   Drain and refill manual transmission/transaxle and final drive unit.   Priority Rating 1

B. Clutch Diagnosis and Repair

B.1.   Diagnose clutch noise, binding, slippage, pulsation, and chatter;
          determine necessary action.   Priority Rating 1

B.2.   Inspect clutch pedal linkage, cables, automatic adjuster mechanisms,
          brackets, bushings, pivots, and springs; perform necessary action.   Priority Rating 1

B.3.   Inspect and replace clutch pressure plate assembly, clutch disc, release
          (throw-out) bearing and linkage, and pilot bearing/bushing
          (as applicable).   Priority Rating 1

B.4.   Bleed clutch hydraulic system.   Priority Rating 1

B.5.   Check and adjust clutch master cylinder fluid level; check for leaks.   Priority Rating 1

B.6.   Inspect flywheel and ring gear for wear and cracks; determine
          necessary action.   Priority Rating 1

B.7.   Measure flywheel runout and crankshaft end play; determine
          necessary action.   Priority Rating 2

C. Transmission/Transaxle Diagnosis and Repair

C.1.   Inspect, adjust, and reinstall shift linkages, brackets, bushings, cables,
          pivots, and levers.   Priority Rating 2

C.2.   Describe the operational characteristics of an electronically controlled
          manual transmission/transaxle.   Priority Rating 3

D. Drive Shaft and Half Shaft, Universal and Constant-Velocity (CV) Joint Diagnosis and Repair

D.1.   Diagnose constant-velocity (CV) joint noise and vibration concerns;
          determine necessary action.   Priority Rating 1

D.2.   Diagnose universal joint noise and vibration concerns; perform
          necessary action.   Priority Rating 2

D.3.   Inspect, remove, and replace front wheel drive (FWD) bearings, hubs,
          and seals   Priority Rating 1

D.4.   Inspect, service, and replace shafts, yokes, boots, and universal/
          CV joints.   Priority Rating 1

D.5.   Check shaft balance and phasing; measure shaft runout; measure
          and adjust driveline angles.   Priority Rating 2

E. Drive Axle Diagnosis and Repair

1.   *Ring and Pinion Gears and Differential Case Assembly*

E.1.1.   Clean and inspect differential housing; check for leaks;
            inspect housing vent.   Priority Rating 2

E.1.2.   Check and adjust differential housing fluid level.   Priority Rating 1

E.1.3.   Drain and refill differential housing.   Priority Rating 1

E.1.4.   Inspect and replace companion flange and pinion seal;
            measure companion flange runout.   Priority Rating 2

2.   *Drive Axles*

E.2.1.   Inspect and replace drive axle wheel studs.   Priority Rating 1

E.2.2.   Remove and replace drive axle shafts.   Priority Rating 1

E.2.3.   Inspect and replace drive axle shaft seals, bearings,
            and retainers.   Priority Rating 2

E.3.4.   Measure drive axle flange runout and shaft end play; determine
            necessary action.   Priority Rating 2

F. Four-Wheel Drive/All-Wheel Drive Component Diagnosis and Repair

F.1.   Inspect, adjust, and repair shifting controls (mechanical, electrical,
          and vacuum), bushings, mounts, levers, and brackets.   Priority Rating 3

F.2.   Inspect front-wheel bearings and locking hubs; perform necessary action(s).   Priority Rating 3

F.3.   Check for leaks at drive assembly seals; check vents; check lube level.   Priority Rating 3

F.4.   Identify concerns related to variations in tire circumference and/or final drive ratios.   Priority Rating 3

Master Automobile Service Technology (MAST) Task List

A. General: Drive Train Diagnosis

A.1.   Identify and interpret drive train concerns; determine necessary action.   Priority Rating 1

A.2.   Research applicable vehicle and service information, fluid type, vehicle service history, service precautions, and technical service bulletins.   Priority Rating 1

A.3.   Check fluid condition; check for leaks; determine necessary action.   Priority Rating 1

A.4.   Drain and refill manual transmission/transaxle and final drive unit.   Priority Rating 1

B. Clutch Diagnosis and Repair

B.1.   Diagnose clutch noise, binding, slippage, pulsation, and chatter; determine necessary action.   Priority Rating 1

B.2.   Inspect clutch pedal linkage, cables, automatic adjuster mechanisms, brackets, bushings, pivots, and springs; perform necessary action.   Priority Rating 1

B.3.   Inspect and replace clutch pressure plate assembly, clutch disc, release (throw-out) bearing and linkage, and pilot bearing/bushing (as applicable).   Priority Rating 1

B.4.   Bleed clutch hydraulic system.   Priority Rating 1

B.5.   Check and adjust clutch master cylinder fluid level; check for leaks.   Priority Rating 1

B.6.   Inspect flywheel and ring gear for wear and cracks; determine necessary action.   Priority Rating 1

B.7.   Measure flywheel runout and crankshaft end play; determine necessary action.   Priority Rating 2

C. Transmission/Transaxle Diagnosis and Repair

C.1.   Inspect, adjust, and reinstall shift linkages, brackets, bushings, cables, pivots, and levers.   Priority Rating 2

C.2.   Describe the operational characteristics of an electronically controlled manual transmission/transaxle.   Priority Rating 3

C.3.   Diagnose noise concerns through the application of transmission/transaxle powerflow principles.   Priority Rating 2

C.4.   Diagnose hard shifting and jumping out of gear concerns; determine necessary action.   Priority Rating 2

C.5.   Diagnose transaxle final drive assembly noise and vibration concerns; determine necessary action.   Priority Rating 3

C.6.   Disassemble, inspect clean, and reassemble internal transmission/transaxle components.   Priority Rating 3

D. Drive Shaft and Half Shaft, Universal and Constant-Velocity (CV) Joint Diagnosis and Repair

D.1.   Diagnose constant-velocity (CV) joint noise and vibration concerns; determine necessary action.   Priority Rating 1

D.2.   Diagnose universal joint noise and vibration concerns; perform necessary action.   Priority Rating 2

D.3.   Inspect, remove, and replace front wheel drive (FWD) bearings, hubs, and seals   Priority Rating 1

D.4.   Inspect, service, and replace shafts, yokes, boots, and universal/CV joints.   Priority Rating 1

D.5.   Check shaft balance and phasing; measure shaft runout; measure and adjust driveline angles.   Priority Rating 2

E. Drive Axle Diagnosis and Repair

1. *Ring and Pinion Gears and Differential Case Assembly*

E.1.1.   Clean and inspect differential housing; check for leaks; inspect housing vent.   Priority Rating 2

E.1.2.   Check and adjust differential housing fluid level.   Priority Rating 1

E.1.3.   Drain and refill differential housing.   Priority Rating 1

E.1.4.   Diagnose noise and vibration concerns; determine necessary action.   Priority Rating 2

E.1.5.   Inspect and replace companion flange and pinion seal; measure companion flange runout.   Priority Rating 2

E.1.6.   Inspect ring gear and measure runout; determine necessary action.   Priority Rating 3

E.1.7.   Remove, inspect, and reinstall drive pinion and ring gear, spacers, sleeves, and bearings.   Priority Rating 3

E.1.8.   Measure and adjust drive pinion depth.   Priority Rating 3

E.1.9.   Measure and adjust drive pinion bearing preload.   Priority Rating 3

E.1.10. Measure and adjust side bearing preload and ring and pinion gear total backlash and backlash variation on a differential carrier assembly (threaded cup or shim types).   Priority Rating 3

E.1.11. Check ring and pinion tooth contact patterns; perform necessary action.   Priority Rating 3

E.1.12. Disassemble, inspect, measure, and adjust or replace differential pinion gears (spiders), shaft, side gears, side bearings, thrust washers, and case.   Priority Rating 3

E.1.13. Reassemble and reinstall differential case assembly; measure runout; determine necessary action.   Priority Rating 3

2. *Limited Slip Differential*

E.2.1.   Diagnose noise, slippage, and chatter concerns; determine necessary action.   Priority Rating 3

E.2.2.   Measure rotating torque; determine necessary action.   Priority Rating 3

3. *Drive Axles*

E.3.1.   Inspect and replace drive axle wheel studs.   Priority Rating 1

E.3.2.   Remove and replace drive axle shafts.   Priority Rating 1

E.3.3.   Inspect and replace drive axle shaft seals, bearings, and retainers.   Priority Rating 2

E.3.4.   Measure drive axle flange runout and shaft end play; determine necessary action.   Priority Rating 2

E.3.5.   Diagnose drive axle shafts, bearings, and seals for noise, vibration, and fluid leakage concerns; determine necessary action.   Priority Rating 2

F. Four-Wheel Drive/All-Wheel Drive Component Diagnosis and Repair

F.1.   Inspect, adjust, and repair shifting controls (mechanical, electrical, and vacuum), bushings, mounts, levers, and brackets.   Priority Rating 3

F.2.   Inspect front-wheel bearings and locking hubs; perform necessary action(s).   Priority Rating 3

F.3.   Check for leaks at drive assembly seals; check vents; check lube level.   Priority Rating 3

F.4. Identify concerns related to variations in tire circumference and/or final drive ratios.   Priority Rating 3

F.5. Diagnose noise, vibration, and unusual steering concerns; determine necessary action.   Priority Rating 3

F.6. Diagnose, test, adjust, and replace electrical/electronic components of four-wheel-drive systems.   Priority Rating 3

F.7. Disassemble, service, and reassemble transfer case and components.   Priority Rating 3

DEFINITION OF TERMS USED IN THE TASK LIST

To clarify the intent of these tasks, NATEF has defined some of the terms used in the task list. For a good understanding of what the task includes, refer to this glossary while reading the task list.

adjust	To bring components to specified operational settings.
align	To restore the proper position of components.
analyze	To assess the condition of a component or system.
assemble (reassemble)	To fit together the components of a device or system.
bleed	To remove air from a closed system.
CAN (Controller Area Network)	CAN is a network protocol (SAE J2284/ISO 15765-4) used to interconnect a network of electronic control modules.
check	To verify a condition by performing an operational or comparative examination.
clean	To rid components of foreign matter for the purpose of reconditioning, repairing, measuring, and reassembling.
confirm	To acknowledge something has happened with firm assurance.
demonstrate	To show or exhibit the knowledge of a theory or procedure.
determine necessary action	Indicates that the diagnostic routine(s) is the primary emphasis of a task. The student is required to perform the diagnostic steps and communicate the diagnostic outcomes and corrective actions required to address the concern or problem. The training program determines the communication method (worksheet, test, verbal communication, or other means deemed appropriate) and whether the corrective procedures for these tasks are actually performed.
diagnose	To identify the cause of a problem.
differentiate	To perceive the difference in or between one thing to other things.
disassemble	To separate a component's parts as a preparation for cleaning, inspection, or service.
fill (refill)	To bring the fluid level to a specified point or volume.
find	To locate a particular problem, such as shorts, grounds, or opens in an electrical circuit.
flush	To internally clean a component or system.
high voltage	Voltages of 50 volts and higher.
identify	To establish the identity of a vehicle or component before service; to determine the nature or degree of a problem.
inspect	(see *check*)
install (reinstall)	To place a component in its proper position in a system.
listen	To use audible clues in the diagnostic process; to hear the customer's description of a problem.
locate	Determine or establish a specific spot or area.

lubricate	To employ the correct procedures and materials in performing the prescribed service.
maintain	To keep something at a specified level, position, rate, etc.
measure	To determine existing dimensions/values for comparison to specifications.
mount	To attach or place a tool or component in the proper position.
network	A system of interconnected electrical modules or devices.
on-board diagnostics (OBD)	Diagnostic protocol that monitors computer inputs and outputs for failures.
perform	To accomplish a procedure in accordance with established methods and standards.
perform necessary action	Indicates that the student is to perform the diagnostic routine(s) and perform the corrective action item. If various scenarios (conditions or situations) are presented in a single task, at least one of the scenarios must be accomplished.
priority ratings	Indicates the minimum percentage of tasks, by area, a program must include in its curriculum in order to be certified in that area.
reassemble	(see *assemble*)
refill	(see *fill*)
remove	To disconnect and separate a component from a system.
repair	To restore a malfunctioning component or system to operating condition.
replace	To exchange a component; to reinstall a component.
research	To make a thorough investigation into a situation or matter.
reset	(see *set*)
select	To choose the correct part or setting during assembly or adjustment.
service	To perform a procedure as specified in the owner's or service manual.
set	To adjust a variable component to a given, usually initial, specification.
test	To verify a condition through the use of meters, gauges, or instruments.
torque	To tighten a fastener to a specified degree or tightness (in a given order or pattern if multiple fasteners are involved on a single component).
verify	To confirm that a problem exists after hearing the customer's concern; or, to confirm the effectiveness of a repair.

CROSS-REFERENCE GUIDES

RST Task	Job Sheet
Shop and Personal Safety	
1.	1
2.	2
3.	2
4.	2
5.	2
6.	2
7.	3
8.	1
9.	2
10.	1
11.	1
12.	1
13.	4, 5
14.	6
15.	7
Tools and Equipment	
1.	8
2.	8
3.	8
4.	8
5.	8
Preparing Vehicle for Service	
1.	9
2.	9
3.	9
4.	9
5.	9
Preparing Vehicle for Customer	
1.	9

MLR Task	Job Sheet
A.1	11
A.2	12
A.3	12

MLR Task	Job Sheet
B.1	16
B.2	16
C.1	18
D.1	24
D.2	25
E.1	26
E.2	27
E.3	27
E.1.1	33
F.1	35
F.2	36

AST Task	Job Sheet
A.1	10
A.2	11
A.3	12
A.4	12
B.1	13
B.2	14
B.3	15
B.4	16
B.5	16
B.6	15
B.7	15
C.1	17
C.2	18
D.1	22
D.2	23
D.3	24
D.4	25
D.5	23
E.1.1	26
E.1.2	27
E.1.3	27
E.1.4	29
E.2.1	33
E.2.2	33
E.2.3	33
E.2.4	33
F.1	34
F.2	35
F.3	36
F.4	37

MAST Task	Job Sheet
A.1	10
A.2	11
A.3	12
A.4	12
B.1	13
B.2	14
B.3	15
B.4	16
B.5	16
B.6	15
B.7	15
C.1	17
C.2	18
C.3	19
C.4	20
C.5	19
C.6	21
D.1	22
D.2	23
D.3	24
D.4	25
D.5	23
E.1.1	26
E.1.2	27
E.1.3	27
E.1.4	28
E.1.5	29
E.1.6	30
E.1.7	30
E.1.8	30
E.1.9	30
E.1.10	30
E.1.11	30
E.1.12	30
E.1.13	31
E.2.1	32
E.2.2	32
E.3.1	33
E.3.2	33
E.3.3	33
E.3.4	33

MAST Task	Job Sheet
E.3.5	33
F.1	34
F.2	35
F.3	36
F.4	37
F.5	38
F.6	39
F.7	40

PREFACE

The automotive service industry continues to change with the technological changes made by automobile, tool, and equipment manufacturers. Today's automotive technician must have a thorough knowledge of automotive systems and components, good computer skills, exceptional communication skills, good reasoning, the ability to read and follow instructions, and above average mechanical aptitude and manual dexterity.

This new edition, like the last, was designed to give students a chance to develop the same skills and gain the same knowledge that today's successful technician has. This edition also reflects the changes in the guidelines established by the National Automotive Technicians Education Foundation (NATEF), in 2013.

The purpose of NATEF is to evaluate technician training programs against standards developed by the automotive industry and recommend qualifying programs for certification (accreditation) by ASE (National Institute for Automotive Service Excellence). Programs can earn ASE certification upon the recommendation of NATEF. NATEF's national standards reflect the skills that students must master. ASE certification through NATEF evaluation ensures that certified training programs meet or exceed industry-recognized, uniform standards of excellence.

At the expense of much time and thought, NATEF has assembled a list of basic tasks for each of their certification areas. These tasks identify the basic skills and knowledge levels that competent technicians have. The tasks also identify what is required for a student to start a successful career as a technician.

In June 2013, after many discussions with the industry, NATEF established a new model for automobile program standards. This new model is reflected in this edition and covers the new standards, which are based on three (3) levels: Maintenance & Light Repair (MLR), Automobile Service Technician (AST), and Master Automobile Service Technician (MAST). Each successive level includes all the tasks of the previous level in addition to new tasks. In other words, the AST task list includes all of the MLR tasks plus additional tasks. The MAST task list includes all of the AST tasks, plus additional tasks designated specifically for MAST.

Most of the content in this book are job sheets. These job sheets relate to the tasks specified by NATEF, according to the appropriate certification level. The main considerations during the creation of these job sheets were student learning and program certification by NATEF. Students are guided through standard industry-accepted procedures. While they are progressing, they are asked to report their findings as well as offer their thoughts on the steps they have just completed. The questions asked of the students are thought provoking and require students to apply what they know to what they observe.

The job sheets were also designed to be generic. That is, whenever possible, the tasks can be performed on any vehicle from any manufacturer. Also, completion of the sheets does not require the use of specific brands of tools and equipment. Rather, students use what is available. In addition, the job sheets can be used as a supplement to any good textbook.

Also included are description and basic use of the tools and equipment listed in NATEF's standards. The standards recognize that not all programs have the same needs, nor do all programs teach all of the NATEF tasks. Therefore, the basic philosophy for the tools and equipment requirement is that the training should be as thorough as possible with the tools and equipment necessary for those tasks.

Theory instruction and hands-on experience of the basic tasks provide initial training for employment in automotive service or further training in any or all of the specialty areas. Competency in the tasks indicates to employers that you are skilled in that area. You need to know the appropriate theory, safety, and support information for each required task. This should include identification and use of the required tools and testing and measurement equipment required for the tasks, the use of current reference and training materials, the proper way to write work orders and warranty reports, and the storage, handling, and use of hazardous materials as required by the "Right to Know" law, and federal, state, and local governments.

Words to the Instructor: We suggest you grade these job sheets based on completion and reasoning. Make sure the students answer all the questions. Then, look at their reasoning to see if the task was actually completed and to get a feel for their understanding of the topic. It will be easy for students to copy others' measurements and findings, but each student should have their own base of understanding and that will be reflected in their explanations.

Words to the Student: While completing the job sheets, you have a chance to develop the skills you need to be successful. When asked for your thoughts or opinions, think about what you observed. Think about what could have caused those results or conditions. You are not being asked to give accurate explanations for everything you do or observe. You are only asked to think. Thinking leads to understanding. Good technicians are good because they have a basic understanding of what they are doing and why they are doing it.

Jack Erjavec and Ken Pickerill

MANUAL TRANSMISSIONS

To prepare you to learn what you should learn from completing the job sheets, some basics must be covered. The discussion begins with an overview of manual transmissions and drivelines, with the emphasis placed on what they do and how they work. This includes the major components and designs of manual transmissions and drivelines and their role in the efficient operation of manual transmissions and drivelines of all designs.

Preparation for working on an automobile is not complete if certain safety issues are not first addressed. This discussion covers what you should and should not do while working on manual transmissions and drivelines, including the proper ways to deal with hazardous and toxic materials.

NATEF's task list for manual transmissions certification is given, along with definitions of some terms used to describe the tasks. The list gives you a good look at what the experts say you need to know before you can be considered competent to work on manual transmissions and drivelines.

Following the task list are descriptions of the various tools and types of equipment you need to be familiar with. These are the tools you will use to complete the job sheets. They are also the tools NATEF has identified as necessary for servicing manual transmissions and drivelines.

After the tool discussion is a cross-reference guide that shows which NATEF tasks are related to specific job sheets. In most cases, there is a single job sheet for each task. Some tasks are part of a procedure, in which case one job sheet may cover two or more tasks. The remainder of the book contains the job sheets.

BASIC MANUAL TRANSMISSION THEORY

Transmissions or transaxles, drive axles, and differentials perform the important task of manipulating the power produced by the engine and routing it to the driving wheels of the vehicle. Precision-machined and fitted gear sets and shafts change the ratio of speed and power between the engine and drive axles. The flow of power is controlled through a manually operated clutch and shift lever.

Clutch

The clutch (Figure 1) is located between the transmission and the engine, where it provides a mechanical coupling between the engine's flywheel and the transmission's input shaft. The driver operates the clutch through a cable or hydraulic system that extends from the passenger

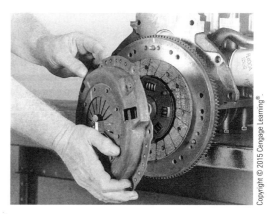

Figure 1 The three main parts of a clutch assembly: the pressure plate, clutch disc, and flywheel.

compartment to the bell housing (also called the clutch housing) between the engine and the transmission.

The clutch engages the transmission gradually by allowing a certain amount of slippage between the transmission's input shaft and the flywheel. The components of a clutch assembly are the flywheel, clutch disc, pressure plate assembly, clutch release bearing (or throwout bearing), and the clutch fork.

The flywheel and the pressure plate are the drive members of the clutch. The drive member connected to the transmission input shaft is the clutch disc, also called the friction disc. As long as the clutch is disengaged (clutch pedal depressed), the drive members turn independently of the driven member, and the engine is disconnected from the transmission. However, when the clutch is engaged (clutch pedal released), the pressure plate moves against the clutch disc, is squeezed between the two revolving drive members, and is forced to turn at the same speed.

The flywheel is normally made of nodular cast iron, which has a high graphite content to lubricate the engagement of the clutch. Welded to or pressed onto the outside diameter of the flywheel is the starter ring gear. The rear surface of the flywheel is a friction surface, machined very flat to ensure smooth clutch engagement. The flywheel absorbs some of the torsional vibrations of the crankshaft and provides the inertia to rotate the crankshaft through the four strokes.

A bore in the center of the flywheel and crankshaft holds the pilot bushing or bearing, which supports the front end of the transmission's input shaft and keeps it aligned with the engine's crankshaft.

The clutch disc receives the driving motion from the flywheel and pressure plate assembly and transmits that motion to the transmission input shaft. Both sides of the clutch disc are covered with friction material, called friction facings.

Grooves are cut across the face of the friction facings to promote clean disengagement of the driven disc from the flywheel and pressure plate; it also promotes better cooling. The facings are riveted to wave springs, also called cushioning springs, which cause the contact pressure on the facings to rise gradually as the springs flatten out when the clutch is engaged. These springs eliminate chatter when the clutch is engaged and reduce the chance of the clutch disc sticking to the flywheel and pressure plate surfaces when

the clutch is disengaged. The wave springs and friction facings are fastened to the steel disc.

The clutch disc is designed to absorb such things as crankshaft vibration, abrupt clutch engagement, and driveline shock. Torsional coil springs allow the disc to rotate slightly in relation to the pressure plate while they absorb the torque forces. The number and tension of these springs is determined by engine torque and vehicle weight. Stop pins limit this torsional movement.

The pressure plate assembly squeezes the clutch disc onto the flywheel with sufficient force to transmit engine torque efficiently. It also must move away from the clutch disc so that the clutch disc can stop rotating, even though the flywheel and pressure plate continue to rotate.

Basically, there are two types of pressure plate assemblies: those with coil springs and those with a diaphragm spring. Both types have a steel cover that bolts to the flywheel and acts as a housing to hold the parts together. In both, there is also the pressure plate, which is a heavy, flat ring made of cast iron. The assemblies differ in the manner in which they move the pressure plate toward and away from the clutch disc.

The pilot bushing or bearing supports the outer end of the transmission's input shaft. This shaft is splined to the clutch disc and transmits power from the engine (when the clutch is engaged) to the transmission. A large bearing in the transmission case supports the transmission end of the input shaft. Because the input shaft extends unsupported from the transmission, a pilot bushing is used to keep it in position. By supporting the shaft, the pilot bushing keeps the clutch disc centered in the pressure plate.

The clutch release bearing, also called a throwout bearing, is usually a sealed, prelubricated ball bearing. Its function is to move the pressure plate release levers or diaphragm spring smoothly and quietly through the engagement and disengagement process.

The release bearing is mounted on an iron casting called a hub, which slides on a hollow shaft at the front of the transmission housing. This hollow shaft is part of the transmission's input shaft bearing retainer.

To disengage the clutch, the release bearing is moved forward on its shaft by the clutch fork. The clutch fork is a forked lever that pivots on a ball stud located in an opening in the bell housing. The forked end slides over the hub of the release bearing and the small end protrudes from the bell housing and connects to the clutch linkage and clutch pedal. The clutch fork moves the release bearing and hub back and forth during engagement and disengagement. Some vehicles use an integrated assembly that combines the slave cylinder with the throwout bearing. This eliminates the need for the clutch fork and pivot.

The driver controls the engagement and disengagement of the clutch assembly smoothly and with little effort, either hydraulically or through the use of a clutch cable. The vast majority of modern vehicles are equipped with hydraulically operated clutches

Often, the clutch assembly is controlled hydraulically. In a hydraulic system, fluid pressure transmits motion from one sealed cylinder to another through a hydraulic line. The hydraulic pressure developed by the master cylinder decreases required pedal effort and provides a precise method of controlling clutch operation.

When the clutch pedal is depressed, a pushrod in the clutch master cylinder moves a piston to create hydraulic pressure. This pressure is passed through a line to the clutch slave cylinder. The pressure moves the piston of the slave. Piston travel is transmitted by a pushrod to the clutch fork. The pushrod boot keeps contaminants out of the slave cylinder.

Transmissions

Transmissions and transaxles serve basically the same purpose and operate by the same basic principles. Although the assembly of a transaxle is different than a transmission, the fundamentals and basic components are the same for both. In fact, a transaxle is basically a transmission with the final drive unit housed within the assembly.

A transmission is a system of gears that transfers the engine's power to the drive wheels of the car. The transmission receives torque from the engine through its input shaft when the clutch is engaged. The torque is then transferred through a set of gears, which either multiply it or transfer it directly. The resultant torque turns the transmission's output shaft, which is indirectly connected to the drive wheels. All transmissions have two primary purposes: they select different speed ratios for a variety of conditions and provide a way to reverse the movement of the vehicle.

Engine torque is applied to the transmission's input shaft when the clutch is engaged. The output shaft (mainshaft) is inserted into the input shaft, but rotates independently of it. The input shaft bearing and a bearing at the rear of the transmission case support the mainshaft. The various speed gears rotate on the mainshaft. Located below or to the side of the input and mainshaft assembly is a countershaft, which is fitted with several sized gears. All these gears, except one, are in constant mesh with the gears on the mainshaft. The remaining gear is in constant mesh with the input gear.

Gear changes occur when a gear is selected by the driver and is locked or connected to the mainshaft. This is accomplished by the movement of a collar that connects the gear to the shaft. Smooth and quiet shifting is possible only when the gears and shaft are rotating at the same speed.

Synchronizers

A synchronizer's primary purpose is to bring components that are rotating at different speeds to one synchronized speed. It also serves to lock these parts together. A single synchronizer is placed between two different speed gears; therefore, transmissions have two or three synchronizer assemblies.

Four types of synchronizers are used in synchromesh transmissions: the block synchronizer, disc and plate synchronizer, plain synchronizer, and pin synchronizer. The most commonly used type on current transmissions is the block type. All synchronizers use friction to synchronize the speed of the gear and shaft before the connection is made.

Block synchronizers consist of a hub, sleeve, blocking ring, and inserts or spring-and-ball detent devices. The synchronizer sleeve surrounds the synchronizer assembly and meshes with the external splines of the hub. The hub is internally splined to the transmission's mainshaft. The outside of the sleeve is grooved to accept the shifting fork. Three slots are equally spaced around the outside of the hub and are fitted with the synchronizer's inserts or spring-and-ball detent assemblies.

The synchronizer's inserts are able to slide back and forth freely in the slots. The inserts are designed with a ridge on their outer surface and insert springs hold this ridge in contact with an internal groove in the synchronizer sleeve. If the synchronizer assembly uses spring-and-ball detents, the balls are held in this groove by their spring. The sleeve is machined to allow it to slide smoothly on the hub.

Bronze or brass blocking rings are positioned at the front and rear of each synchronizer assembly. Blocking rings have notches to accept the insert keys that cause it to rotate at the same speed as the hubs. Around the outside of the blocking ring is a set of beveled dog teeth. These teeth are used for alignment during the shift sequence. The inside of the blocking ring is shaped like a cone, the surface of which has many sharp grooves. These inner surfaces of the blocking rings match the conical shape of the shoulders of the driven gear. These cone-shaped surfaces serve as the frictional surfaces for the synchronizer. The shoulder of the gear also has a ring of beveled dog teeth designed to align with the dog teeth on the blocking ring.

Transmission Designs

All automotive transmissions and transaxles are equipped with a varied number of forward speed gears and one reverse speed. Five-speed transmissions and transaxles are now the most commonly used units. Some of the earlier units were actually four-speeds with an add-on fifth or overdrive gear. Most late-model units incorporate fifth gear in their main assemblies. This is also true of late-model six-speed transmissions and transaxles. The fifth and sixth gears of these units are part of the main gear assembly. Most often each of the two high gears in a six-speed provides an overdrive gear.

Basic Operation

All manual transmissions function in much the same way and have similar parts. The transmission case houses most of the gears in the transmission and has machined surfaces for a cover, rear extension housing, and mounting to the clutch's bell housing. The front bearing retainer is a cast-iron piece bolted to the front of the transmission case.

The counter gear assembly is normally located in the lower portion of the transmission case and is constantly meshed with the input gear. The countershaft contains several different-sized gears that rotate as one solid assembly. Normally the counter gear assembly rotates on several rows of roller or needle bearings.

The speed gears are located on the mainshaft but are not fastened to it and rotate freely on the shaft's journals. The gears are constantly in mesh with the counter gears. The outer end of the mainshaft has splines for the slip-joint yoke of the driveshaft.

A reverse gear is not meshed with the counter gear, as the forward gears are; rather, the reverse idler gear is meshed with it. Normally, reverse gear is engaged by sliding it into mesh with the reverse idler gear. The addition of this third gear causes the reverse gear to rotate in the direction opposite to that of the forward gears.

Shift forks move the synchronizer sleeves to engage and disengage gears. Most shift forks have two fingers that ride in the groove on the outside of the sleeve. The forks are bored to fit over shift rails. Tapered pins are commonly used to fasten the shift forks to the rails. Each of the shift rails has shift lugs that the shift lever fits into. These lugs are also fastened to the rails by tapered pins. The shift rails slide back and forth in bores of the transmission case. Each shift rail has notches in which a spring-loaded ball or bullet drops in to give a detent feel to the shift lever and locate the proper position of the shift fork during gear changes. The shift rails also have notches cut in their sides for interlock plates or pins to fit in. Interlock plates prevent the engagement of two gears at the same time. The lower portion of the shifter assembly fits into the shift rail lugs and moves the shift rails for gear selection.

The ends of the shafts are fitted with large roller or ball bearings pressed onto the shaft. Some transmissions with long shafts use an intermediate bearing to give added strength to the shaft. Small roller or needle bearings are often used on the countershaft, reverse idler gear shaft, and at the connection of the output shaft to the input shaft.

Gear Shift Linkages

Gear shift linkages are either internal or external to the transmission, although the majority of modern transmissions use internal linkages. Internal linkages are located at the side or top of the transmission housing. The control end of the shifter is mounted inside the transmission, as are all the shift controls. Movement of the shifter moves a shift rail and shift fork toward the desired gear and moves the synchronizer sleeve to lock the speed gear to the shaft.

External linkages use rods that are external to the transmission to act on levers connected to the transmission's internal shift rails.

Transaxles

The transmission section of a transaxle is practically identical to that of rear-wheel-drive (RWD) transmissions. It provides for torque multiplication, allows for gear shifting, and is synchronized. It also uses many of the design and operating principles found in transmissions. However, a transaxle also contains the differential gear sets and the connections for the drive axles.

One of the primary distinctions of a transaxle is the absence of the cluster gear assembly. The input shaft gears drive the output shaft gears directly. The output shaft rotates according to each synchro-activated gear's operating ratio.

Normally, a transaxle has two separate shafts: an input shaft and an output shaft. Engine torque is applied to the input shaft, and the revised torque (due to the transaxle's gearing) rotates the output shaft. Normally, the input shaft is located above and parallel to the output shaft. The main speed gears freewheel around the output shaft unless they are locked to the shaft by synchronizers. The main speed gears are in constant mesh with their mating gears on the input shaft and rotate whenever the input shaft rotates.

Some transaxles are equipped with an additional shaft designed to offset the power flow on the output shaft. Power is transferred from the output shaft to the third shaft. The third shaft is added only when an extremely compact transaxle installation is required. The third shaft in some transaxles is an offset input shaft that receives the engine's power and transmits it to a mainshaft.

Some transaxles use a two-cable assembly to shift the gears. One cable is the transmission selector cable and the other is a shifter cable. The selector cable activates the desired shift fork, and the shifter cable causes the engagement of the desired gear.

A pinion gear is machined onto the end of the transaxle's output shaft. This pinion gear is in constant mesh with the differential ring gear. When the output shaft rotates, the pinion gear causes the ring gear to rotate. The resultant torque rotates the other differential gears, which in turn rotate the vehicle's drive axles and wheels.

Front-Wheel-Drive Final Drive Units

One major difference between the differential in a RWD car and the differential in a transaxle is power flow. In a RWD differential, the power flow changes 90 degrees between the drive pinion gear and the ring gear. This change in direction is not needed with most front-wheel-drive (FWD) cars. The transverse engine position places the crankshaft so that it is already rotating in the correct direction. Therefore, the purpose of the differential is only to provide torque multiplication and divide the torque between the drive axle shafts so that they can rotate at different speeds.

Four common configurations are used as the final drives on FWD vehicles: helical, planetary, hypoid, and chain drive. The helical, planetary, and chain final drive arrangements are usually found in transversely mounted power trains. Hypoid final drive gear assemblies are normally used with longitudinal powertrain arrangements.

Front-Wheel-Drive Drive Axles

The differential side gears are connected to inboard constant velocity (CV) joints by splines. The drive axles extend out from each side of the differential to rotate the car's wheels. The axles are made up of three pieces, linked together, to allow the wheels to turn for steering and to move up and down with the suspension. A short stub shaft extends from the differential to the inner CV joint. Connecting the inner CV joint and the outer CV joint is an axle shaft. Extending from the outer CV joint is a short spindle shaft that fits into the hub of the wheels. To keep dirt and moisture out of the CV joints, a neoprene boot is installed over each CV joint assembly.

Rear-Wheel-Drive Axle Shafts and Bearings

RWD vehicles use a single housing to mount the final drive gears and axles. The entire housing is part of the suspension and helps to locate the rear wheels. Located within the hollow horizontal tubes of the axle housing are the axle shafts. Axle shafts are heavy steel bars splined at the inner end to mesh with the axle side gear in the differential. The driving wheel is bolted to the wheel flange at the outer end of the axle shaft.

There are basically three designs of axles: full-floating, three-quarter-floating, and semifloating. The names refer to where the axle bearing is placed in relation to the axle and the housing and how the axle is supported. The bearing of a full-floating axle is located on the outside of the housing. This places the entire vehicle's weight on the axle housing and no weight on the axle.

Three-quarter-floating and semifloating axles are supported by bearings located in the housing and thereby carry some of the weight of the vehicle. Most passenger cars are equipped with three-quarter-floating or semifloating axles. Full-floating axles are commonly found on heavy-duty trucks.

Three designs of axle shaft bearings are used on semifloating axles: ball-type, straight-roller, and tapered-roller bearings. The end-to-end movement of the axle is controlled by a C-type retainer on the inner end of the axle shaft or by a bearing retainer and retainer plate at the outer end of the axle shaft.

The drive axles on most new independent rear suspension (IRS) systems use two U- or CV joints per axle to connect the axle to the differential and the wheels. They are also equipped with linkages and control arms to limit camber changes. The axles of an IRS system are much like those of a FWD system. The outer portion of the axle is supported by an upright or locating member, which is also part of the suspension.

Differentials

A differential is a geared mechanism located between the two driving axles. It rotates the driving axles at different speeds when the vehicle is turning a corner. It also allows both axles to turn at the same speed when the vehicle is moving straight. The gear ratio of the differential's ring-and-pinion gear is used to increase torque, which improves drivability.

On RWD vehicles, power from the transmission's output shaft is transferred by the vehicle's drive shaft to the axle assembly through the pinion flange. This flange is the connecting yoke to the rear universal joint. Power then enters the differential assembly on the pinion gear. The pinion teeth engage the ring gear, which is mounted upright at a 90-degree angle to the pinion. Therefore, as the driveshaft turns, so do the pinion and ring gears.

The ring gear is fastened to the differential case, which is supported by two tapered roller bearings and the rear axle housing. Two beveled differential pinion gears and thrust washers are mounted on the differential pinion shaft. Meshed with the differential pinion gears are two axle side gears splined internally to mesh with the external splines on the left and right axle shafts. The pinion gears and side gears form a square of gears inside the differential case.

The differential pinion gears are free to rotate on their own centers and can travel in a circle as the differential case and pinion shaft rotate. The side gears are meshed with the pinion gears and are free to rotate on their own centers. However, since the side gears are mounted at the centerline of the differential case, they do not travel in a circle with the differential case.

When a car is moving straight ahead, both drive wheels are able to rotate at the same speed. Engine power comes in on the pinion gear and rotates the ring gear. The ring gear rotates the differential case and carries around the pinion gears. As a result, all the gears rotate as a single unit. Each side gear rotates at the same speed and in the same plane, as does the case. Each wheel rotates at the same speed because each axle receives the same rotation.

As the vehicle goes around a corner, the inside wheel travels a shorter distance than the outside wheel. The inside wheel must therefore rotate slower than the outside wheel. In this situation, the differential pinion gears will "walk" forward on the slower-turning or inside side gear. As the pinion gears walk around the slower side gear, they drive the other side gear at a greater speed. An equal percentage of speed is removed from one axle and given to the other; however, the torque applied to each wheel is equal.

Drive Shafts and Universal Joints

On front-engine RWD cars, the torque at the transmission's output shaft is carried through a driveshaft to the final drive unit.

The driveshaft is a tube with universal joint yokes welded to both ends of it. Some drivelines have two driveshafts and four universal joints and use a center support bearing, which serves as the connecting link between the two halves. Four-wheel-drive (4WD) vehicles use two driveshafts: one to drive the front wheels and the other to drive the rear wheels.

As the rear wheels encounter irregularities in the road, the vehicle's springs compress or expand. This changes the angle of the driveline between the transmission and the rear axle housing. It also changes the distance between the transmission and the differential.

In order for the driveshaft to respond to these constant changes, it is equipped with one or more universal joints, which permit variations in the angle of the shaft, and a slip joint, which permits the effective length of the driveline to change.

A universal joint is basically a double-hinged joint consisting of two Y-shaped yokes, one on the driving or input shaft and the other on the driven or output shaft, plus a cross-shaped unit called the cross. A yoke is used to connect the U-joint to the shaft. The four arms of the cross are fitted with bearings in the ends of the two shaft yokes. The input shaft's yoke causes the cross to rotate, and the two other trunnions of the cross cause the output shaft to rotate. When the two shafts are at an angle to each other, the bearings allow the yokes to swing around on their trunnions with each revolution. This action allows two shafts, at a slight angle to each other, to rotate together.

Universal joints allow the driveshaft to transmit power to the rear axle through varying angles that result from the travel of the rear suspension. There are three common designs of universal joints: single universal joints, retained by either an inside or outside snap ring; coupled universal joints; and universal joints held in the yoke by U-bolts or lock plates.

The slip joint assembly includes the transmission's output shaft, the slip joint itself, a yoke, U-joint, and the driveshaft. The output shaft has external splines, which match the internal splines of the slip joint. The meshing of the splines allows the two shafts to rotate together but permits the ends of the shafts to slide along each other. This sliding motion allows for an effective change in the length of the driveshaft, as the drive axles move toward or away from the car's frame. A U-joint connects the yoke of the slip joint to the driveshaft.

Four-Wheel-Drive Systems

Four-wheel-drive (4WD) vehicles designed for off-the-road use are normally RWD vehicles equipped with a transfer case, a front driveshaft, and a front differential and drive axles. Many 4WD vehicles use three driveshafts. One short driveshaft connects the output of the transmission

to the transfer case. The output from the transfer case is then sent to the front and rear axles through separate driveshafts.

Some high-performance cars are equipped with 4WD to improve the handling characteristics of the car. Many of these cars are FWD models converted to 4WD. Normally, FWD cars are modified by adding a transfer case, a rear driveshaft, and a rear axle with a differential. Although this is the typical modification, some cars are equipped with a center differential in place of the transfer case. This differential unit allows the rear and front wheels to turn at different speeds and with different amounts of torque.

The transfer case is usually mounted to the side or rear of the transmission. When a driveshaft is not used to connect the transmission to the transfer case, a chain or gear drive, within the transfer case, receives the engine's power from the transmission, and transfers it to the driveshafts leading to the front and rear drive axles.

The transfer case itself is constructed similar to a transmission. It uses shift forks to select the operating mode, as well as splines, gears, shims, bearings, and other components found in transmissions. The housing is filled with lubricant that reduces friction on all moving parts. Seals hold the lubricant in the case and prevent leakage around shafts and yokes. Shims set up the proper clearance between the internal components and the case.

An electric switch or shift lever, located in the passenger compartment, controls the transfer case so that power is directed to the axles selected by the driver. Power can typically be directed to all four wheels, two wheels, or none of the wheels. On many vehicles, the driver can also select a low-speed range for extra torque while traveling in very adverse conditions.

The rear drive axle of a 4WD vehicle is identical to those used in two-wheel-drive vehicles. The front drive axle is also like a conventional rear axle, except that it is modified to allow the front wheels to steer. Further modifications are also necessary to adapt the axle to the vehicle's suspension system. The differential units housed in the axle assemblies are similar to those found in a RWD vehicle.

Hybrid Vehicles

A hybrid vehicle has at least two different types of power or propulsion systems. Today's hybrid vehicles have an internal combustion engine and an electric motor (Some vehicles have more than one electric motor.). A hybrid's electric motor is powered by high-voltage batteries and/or ultra-capacitors, which are recharged by a generator that is driven by the engine. They are also recharged through regenerative braking. The engine may use gasoline, diesel, or an alternative fuel. Complex electronic controls monitor the operation of the vehicle. Based on the current operating conditions, electronics control the engine, electric motor, and generator.

Depending on the design of the hybrid vehicle, the engine may power the vehicle, assist the electric motor while it is propelling the vehicle, or drive a generator to charge the vehicle's batteries. The electric motor may propel the vehicle by itself, assist the engine while it is propelling the vehicle, or act as a generator to charge the batteries. Many hybrids rely exclusively on the electric motor(s) during slow speed operation, the engine at higher speeds, and both during some certain driving conditions.

SAFETY

In an automotive repair shop, there is great potential for serious accidents, simply because of the nature of the business and the equipment used. Through carelessness, the automotive repair industry can be one of the most dangerous occupations. However, the chances of being injured while working on a car are close to nil if you learn to work safely and use common sense. Safety is the responsibility of everyone in the shop.

Personal Protection

Some procedures, such as grinding, result in tiny particles of metal and dust being thrown off at very high speeds. These metal and dirt particles can easily get into your eyes, causing scratches or cuts on your eyeball. Pressurized gases and liquids escaping a ruptured hose or hose fitting can spray a great distance. If these chemicals get into your eyes, they can cause blindness. Dirt and sharp bits of corroded metal can easily fall into your eyes while you are working under a vehicle.

Eye protection should be worn whenever you are exposed to these risks. To be safe, you should wear safety glasses whenever you are working in the shop. Some procedures may require that you wear other eye protection in addition to safety glasses; for example, when cleaning parts with a pressurized spray, you should wear a face shield. The face shield not only gives added protection to your eyes, but it also protects the rest of your face.

If chemicals such as battery acid, fuel, or solvents get into your eyes, flush them continuously with clean water. Have someone call a doctor, and get medical help immediately.

Your clothing should be well fitted and comfortable but made with strong material. Loose, baggy clothing can easily get caught in moving parts and machinery. Some technicians prefer to wear coveralls or shop coats to protect their personal clothing. Your work clothing should offer you some protection but should not restrict your movement.

Long hair and loose, hanging jewelry can create the same type of hazard as loose-fitting clothing. They can get caught in moving engine parts and machinery. If you have long hair, tie it back or tuck it under a cap.

Never wear rings, watches, bracelets, or neck chains. These can easily get caught in moving parts and cause serious injury.

Always wear leather or similar material shoes or boots with non-slip soles. Steel-tipped safety shoes can give added protection to your feet. Jogging or basketball shoes, street shoes, and sandals are inappropriate in the shop.

Good hand protection is often overlooked. A scrape, cut, or burn can limit your effectiveness at work for many days. A well-fitted pair of heavy work gloves should be worn during operations such as grinding and welding or when handling high-temperature components. Always wear approved rubber gloves when handling strong and dangerous caustic chemicals.

Many technicians wear thin, surgical-type latex gloves whenever they are working on vehicles. These offer little protection against cuts but do offer protection against disease and grease buildup under and around your fingernails. These gloves are comfortable and are quite inexpensive.

Accidents can be prevented simply by the way you act. Following are some guidelines for working in a shop. This list does not include everything you should or shouldn't do; it merely gives some things to think about.

- Never smoke while working on a vehicle or while working with any machine in the shop.

- Playing around is not fun when it sends someone to the hospital.

- To prevent serious burns, keep your skin away from hot metal parts such as the radiator, exhaust manifold, tailpipe, catalytic converter, and muffler.

- Always disconnect electric engine cooling fans when working around the radiator. Many of these can turn on without warning and can easily chop off a finger or hand. Make sure you reconnect the fan after you have completed your repairs.

- When working with a hydraulic press, make sure the pressure is applied in a safe manner. It is generally wise to stand to the side when operating the press.

- Properly store all parts and tools by putting them away in a place where people will not trip over them. This practice not only cuts down on injuries, it also reduces time wasted looking for a misplaced part or tool.

Work Area Safety

Your entire work area should be kept clean and safe. Any oil, coolant, or grease on the floor can make it slippery. To clean up oil, use commercial oil absorbent. Keep all water off the floor. Water is slippery on smooth floors, and electricity flows well through water. Aisles and walkways should be kept clean and wide enough to move through easily. Make sure the work areas around machines are large enough to operate machines safely.

Gasoline is a highly flammable volatile liquid. Something that is *flammable* catches fire and burns easily. A *volatile* liquid is one that vaporizes very quickly. *Flammable volatile* liquids are potential firebombs. Always keep gasoline or diesel fuel in an approved safety can, and never use gasoline to clean your hands or tools.

Handle all solvents (and any liquids) with care to avoid spillage. Keep all solvent containers closed, except when pouring. Proper ventilation is very important in areas where volatile solvents and chemicals are used. Solvent and other combustible materials must be stored in approved and designated storage cabinets or rooms with adequate ventilation. Never light matches or smoke near flammable solvents and chemicals, including battery acids.

Oily rags should also be stored in an approved metal container. When these oily, greasy, or paint-soaked rags are left lying about or are not stored properly, they can spontaneously combust. *Spontaneous combustion* refers to fire that starts by itself, without a match.

Disconnecting the vehicle's battery before working on the electrical system or before welding can prevent fires caused by a vehicle's electrical system. To disconnect the battery, remove the negative or ground cable from the battery and position it away from the battery.

Know where all the shop's fire extinguishers are located. Fire extinguishers are clearly labeled as to type and types of fire they should be used on. Make sure you use the correct type of extinguisher for the type of fire you are dealing with. A multipurpose dry chemical fire extinguisher puts out ordinary combustibles, flammable liquids, and electrical fires. Never put water on a gasoline fire. The water will just spread the fire—the proper fire extinguisher smothers the flames.

During a fire, never open doors or windows unless it is absolutely necessary; the extra draft will only make the fire worse. Make sure the fire department is contacted before or during your attempt to extinguish a fire.

The storage batteries of hybrid vehicles have enough power to easily KILL you if they are not disconnected/disarmed before starting work on the vehicle. Make sure you are trained in the proper procedures and have the necessary personal protective gear, such as special gloves and shoes.

Tool and Equipment Safety

Careless use of simple hand tools, such as wrenches, screwdrivers, and hammers, causes many shop accidents that could be prevented. Keep all hand tools free of grease and in good condition. Tools that slip can cause cuts and bruises. If a tool slips and falls into a moving part, it can fly out and cause serious injury.

Use the proper tool for the job, and make sure the tool is of professional quality. Using poorly made tools or the wrong tools can damage parts, the tool itself, or you. Never use broken or damaged tools.

Safety around power tools is very important. Serious injury can result from carelessness. Always wear safety glasses when using power tools.

If the tool is electrically powered, make sure it is properly grounded. Before using it, check for bare wires or cracks in the insulation. When using electrical power tools, never stand on a wet or damp floor. Never leave a running power tool unattended.

Tools that use compressed air are called *pneumatic tools*. Compressed air is used to inflate tires, apply paint, and drive tools. Compressed air can be dangerous when it is not used properly.

When using compressed air, wear safety glasses or a face shield, or both. Particles of dirt and pieces of metal blown by the high-pressure air can penetrate your skin or get into your eyes.

Before using a compressed air tool, check all hose connections. Always hold an air nozzle or air control device securely when starting or shutting off the compressed air. A loose nozzle can whip suddenly and cause serious injury. Never point an air nozzle at anyone. Never use compressed air to blow dirt from your clothes or hair. Never use compressed air to clean the floor or workbench.

Always be careful when raising a vehicle on a lift or a hoist. Adapters and hoist plates must be positioned correctly to prevent damage to the underbody of the vehicle. There are specific lift points that allow the weight of the vehicle to be supported evenly by the adapters or hoist plates. The correct lift points can be found in the vehicle's service manual. Before operating any lift or hoist, carefully read the operating manual and follow the operating instructions.

Once you know the lift supports are properly positioned under the vehicle, raise the lift until the supports contact the vehicle. Check the supports to make sure they are in full contact with the vehicle. Shake the vehicle to make sure it is securely balanced on the lift, and then raise the lift to the desired working height. Before working under a car, make sure the lift's locking devices are engaged.

A vehicle can be raised off the ground by a hydraulic jack. The jack's lifting pad must be positioned under an area of the vehicle's frame or at one of the manufacturer's recommended lift points. Never place the pad under the floor pan or under steering and suspension components, which are easily damaged by the weight of the vehicle. Always position the jack so the wheels of the vehicle can roll as the vehicle is being raised.

Safety stands, also called jack stands, should be placed under a sturdy chassis member, such as the frame or axle housing, to support the vehicle after it has been raised by a jack. Once the safety stands are in position, the hydraulic pressure in the jack should be released slowly until the weight of the vehicle is on the stands. Never move under a vehicle that is supported only by a hydraulic jack. Rest the vehicle on the safety stands before moving under the vehicle.

Heavy parts of the automobile, such as engines, are removed with chain hoists or cranes. Cranes often are called cherry pickers. To prevent serious injury, chain hoists and cranes must be properly attached to the parts being lifted. Always use bolts with enough strength to support the object being lifted. After you have attached the lifting chain or cable to the part that is being removed, have your instructor check it. Place the chain hoist or crane directly over the assembly, then attach the chain or cable to the hoist.

Cleaning parts is a necessary step in most repair procedures. Always wear the appropriate protection when using chemical, abrasive, and thermal cleaners.

Working Safely on High-Voltage Systems

Electric drive vehicles (battery operated, hybrid, and fuel cell electric vehicles) have high-voltage electrical systems (from 42 volts to 650 volts). These high voltages can kill you! Fortunately, most high-voltage circuits are identifiable by size and color. The cables have thicker insulation and are typically colored orange. The connectors are also colored orange. On some vehicles, the high-voltage cables are enclosed in an orange shielding or casing, again the orange indicates high voltage. In addition, the high-voltage battery pack and most high-voltage components have "High Voltage" caution labels. Be careful not to touch these wires and parts.

Wear insulating gloves, commonly called "lineman's gloves", when working on or around the high-voltage system. These gloves must be class "0" rubber insulating gloves, rated at 1000-volts. Also, to protect the integrity of the insulating gloves, as well as you, wear leather gloves over the insulating gloves while doing a service.

Make sure they have no tears, holes or cracks and that they are dry. Electrons are very small and can enter through the smallest of holes in your gloves. The integrity of the gloves should be checked before using them. To check the condition of the gloves, blow enough air into each one so they balloon out. Then fold the open end over to seal the air in. Continue to slowly fold that end of the glove toward the fingers. This will compress the air. If the glove continues to balloon as the air is compressed, it has no leaks. If any air leaks out, the glove should be discarded. All gloves, new and old, should be checked before they are used.

There are other safety precautions that should always be adhered to when working on an electric drive vehicle:

- Always adhere to the safety guidelines given by the vehicle's manufacture.

- Obtain the necessary training before working on these vehicles.

- Be sure to perform each repair operation following the test procedures defined by the manufacturer.

- Disable or disconnect the high-voltage system before performing services to those systems. Do this according to the procedures given by the manufacturer.

- Anytime the engine is running in a hybrid vehicle, the generator is producing high voltage and care must be taken to prevent being shocked.

- Before doing any service to an electric drive vehicle, make sure the power to the electric motor is disconnected or disabled.

- Systems may have a high-voltage capacitor that must be discharged after the high-voltage system has been isolated. Make sure to wait the prescribed amount of time (normally about 10 minutes) before working on or around the high-voltage system.

- After removing a high-voltage cable, cover the terminal with vinyl electrical tape.

- Always use insulated tools.

- Alert other technicians that you are working on the high-voltage systems with a warning sign such as "High-Voltage Work! Do Not Touch."

- Always install the correct type of circuit protection device into a high-voltage circuit.

- Many electric motors have a strong permanent magnet in them; individuals with a pacemaker should not handle these parts.

- When an electric drive vehicle needs to be towed into the shop for repairs, make sure it is not towed on its drive wheels. Doing this will drive the generator(s), which can overcharge the batteries and cause them to explode.

- Always tow these vehicles with the drive wheels off the ground or move them on a flat bed.

If the electric motor is sandwiched between the engine and transmission, make sure you follow all procedures. The permanent magnet used in the motor is very strong and requires special tools to remove and install it.

Vehicle Operation

When a customer brings a vehicle in for service, shop personnel should follow certain driving rules to ensure the safety of everyone in the shop. For example, before moving a car into the shop, buckle your safety belt. Make sure that no one is nearby, that the way is clear, and that there are no tools or parts under the car before you start the engine.

Check the brakes before putting the vehicle in gear. Then, drive slowly and carefully in and around the shop.

If the engine must be running while work is done on the car, block the wheels to prevent the car from moving. Place the transmission in park for automatic transmissions or in neutral for manual transmissions. Set the parking (emergency) brake. Never stand directly in front of or behind a running vehicle.

Run the engine only in a well-ventilated area to avoid the danger of poisonous carbon monoxide (CO) in the engine exhaust. CO is an odorless but deadly gas. Most shops have an exhaust ventilation system; always use it. Connect the hose from the vehicle's tailpipe to the intake for the vent system. Make sure the vent system is turned on before running the engine. If the work area does not have an exhaust venting system, use a hose to direct the exhaust out of the building.

HAZARDOUS MATERIALS AND WASTES

A typical shop contains many potential health hazards for those working in it. These hazards can cause injury, sickness, impairment, discomfort, and even death. Here is a short list of the different classes of hazards.

- Chemical hazards are caused by high concentrations of vapors, gases, or solids (in the form of dust).
- Hazardous wastes are substances that are the result of a service.
- Physical hazards include excessive noise, vibration, pressures, and temperatures.
- Ergonomic hazards are conditions that impede normal or proper body position and motion.

There are many government agencies charged with ensuring safe work environments for all workers. These include the Occupational Safety and Health Administration (OSHA), Mine Safety and Health Administration (MSHA), and National Institute for Occupational Safety and Health (NIOSH). These agencies, in addition to state and local governments, have instituted regulations that must be understood and followed. Everyone in a shop is responsible for adhering to these regulations.

An important part of a safe work environment is the employees' knowledge of potential hazards. Right-to-know laws concerning all chemicals protect every employee in the shop. The general intent of right-to-know laws is for employers to provide their employees with a safe working place as it relates to hazardous materials.

All employees must be trained about their rights under the legislation, the nature of the hazardous chemicals in the workplace, and the contents of the labels on the chemicals. All information about each chemical must be posted on Material Safety Data Sheets (MSDS) and must be accessible. The manufacturer of the chemical must give these sheets to its customers on request. They detail the chemical composition and precautionary information for all products that can present a health or safety hazard.

Employees must become familiar with the general uses, protective equipment, accident or spill procedures, and any other information about the safe handling of the hazardous material.

This training must be given to employees annually and to new employees as part of their job orientation.

All hazardous material must be properly labeled, indicating what health, fire, or reactivity hazard it poses and what protective equipment is necessary when handling each chemical. The manufacturer of the hazardous materials must provide all warnings and precautionary information, which the user must read and understand before using the material. A list of all hazardous materials used in the shop must be posted for employees to see.

Shops must maintain documentation on the hazardous chemicals in the workplace, proof of training programs, records of accidents or spill incidents, and satisfaction of employee requests for specific chemical information via the MSDS. A general right-to-know compliance procedure manual must be used in the shop.

When handling any hazardous materials or hazardous waste, make sure you follow the required procedures for handling such material. Wear the proper safety equipment listed on the MSDS, which includes the use of approved respirator equipment.

Some of the common hazardous materials that automotive technicians use are cleaning chemicals, fuels (gasoline and diesel), paints and thinners, battery electrolyte (acid), used engine oil, refrigerants, and engine coolant (antifreeze).

Many repair and service procedures generate what are known as hazardous wastes. Dirty solvents and cleaners are good examples of hazardous wastes. A material is classified as a *hazardous waste* if it is on the Environmental Protection Agency (EPA) list of known harmful materials or has one or more of the following characteristics.

- *Ignitability.* A liquid with a flash point below 140°F or a solid that can spontaneously ignite.
- *Corrosivity.* It dissolves metals and other materials or burns the skin.
- *Reactivity.* Any material that reacts violently with water or other materials or releases cyanide gas, hydrogen sulfide gas, or similar gases when exposed to low-pH acid solutions. Included are any materials that generate toxic mists, fumes, vapors, or flammable gases.
- *EP toxicity.* Materials that leach one or more of eight heavy metals in concentrations greater than 100 times primary drinking water standard concentrations.

Complete EPA lists of hazardous wastes can be found in the Code of Federal Regulations. It should be noted that no material is considered hazardous waste until the shop is finished using it and ready to dispose of it.

The following list describes the recommended procedure for dealing with some of the common hazardous wastes. Always follow these and any other mandated procedures.

Oil Recycle oil. Set up equipment, such as a drip table or screen table with a used oil collection bucket, to collect oils dripping off parts. Place drip pans underneath vehicles that are leaking fluids onto the storage area. Do not mix other wastes with used oil, except as allowed by your recycler. Used oil generated by a shop (or oil received from household "do-it-yourself" generators) may be burned on site in a commercial space heater. Used oil also may be burned for energy recovery. Contact state and local authorities to determine requirements and to obtain necessary permits.

Oil filters Drain for at least 24 hours, crush, and recycle used oil filters.

Batteries Recycle batteries by sending them to a reclaimer or back to the distributor. Keeping shipping receipts can demonstrate that you have recycled. Store batteries in a watertight, acid-resistant container. Inspect batteries for cracks and leaks when they come in. Treat a dropped battery as if it were cracked. Acid residue is hazardous because it is corrosive and may contain lead and other toxics. Neutralize spilled acid by using baking soda or lime, and dispose of as hazardous material.

Metal residue from machining Collect metal filings when machining metal parts. Keep separate and recycle if possible. Prevent metal filings from falling into a storm sewer drain.

Refrigerants Recover or recycle refrigerants (or do both) during the service and disposal of motor vehicle air conditioners and refrigeration equipment. It is not allowable to knowingly vent refrigerants to the atmosphere. Recovering or recycling during servicing must be performed by an EPA-certified technician using certified equipment and following specified procedures.

Solvents Replace hazardous chemicals with less toxic alternatives that have equal performance. For example, substitute water-based cleaning solvents for petroleum-based solvent degreasers. To reduce the amount of solvent used when cleaning parts, use a two-stage process (i.e., dirty solvent followed by fresh solvent). Hire a hazardous waste management service to clean and recycle solvents. (Some spent solvents must be disposed of as hazardous waste, unless recycled properly.) Store solvents in closed containers to prevent evaporation. Evaporation of solvents contributes to ozone depletion and smog formation. In addition, the residue from evaporation must be treated as a hazardous waste. Properly label spent solvents and store in drip pans or in diked areas and only with compatible materials.

Containers Cap, label, cover, and properly store above ground and outdoors any liquid containers and small tanks within a diked area and on a paved impermeable surface to prevent spills from running into surface or ground water.

Other solids Store materials such as scrap metal, old machine parts, and worn tires under a roof or tarpaulin to protect them from the elements and to prevent potentially contaminated runoff. Consider recycling tires by retreading them.

Liquid recycling Collect and recycle coolants from radiators. Store transmission fluids, brake fluids, and solvents containing chlorinated hydrocarbons separately, and recycle or dispose of them properly.

Shop towels or rags Keep waste towels in a closed container marked "Contaminated Shop Towels Only." To reduce costs and liabilities associated with disposal of used towels, which can be classified as hazardous wastes, investigate using a laundry service that is able to treat the wastewater generated from cleaning the towels.

Waste storage Always keep hazardous waste separate, properly labeled, and sealed in the recommended containers. The storage area should be covered and may need to be fenced and locked if vandalism could be a problem. Select a licensed hazardous waste hauler after seeking recommendations and reviewing the firm's permits and authorizations.

MANUAL TRANSMISSION TOOLS AND EQUIPMENT

Many different tools and pieces of testing and measuring equipment are used to service transmissions and drivelines. NATEF has identified many of these and determined that a certified manual drive train and axle technician must know what they are and how and when to use them. The tools and equipment listed by NATEF are covered in the following discussion. Also included are the tools and equipment you will use while completing the job sheets. Although you must know and be able to use common hand tools, they are not part of this discussion. You should already know what they are and how to use and care for them.

Portable Crane

To remove and install a transmission, the engine is often moved out of the vehicle with the transmission. To remove or install an engine, a portable crane, frequently called a cherry picker, is used. A crane uses hydraulic pressure that is converted to a mechanical advantage and lifts the engine from the vehicle. To lift an engine, attach a pulling sling or chain to the engine. Some engines have eye plates for use in lifting (Figure 2). If they are not available, the sling must be bolted to the engine. The sling-attaching bolts must be large enough to support the engine and must thread into the block a minimum of 1.5 times the bolt diameter. Connect the crane to the chain. Raise the engine slightly and make sure the sling attachments are secure. Carefully lift the engine out of its compartment.

Lower the engine close to the floor so the transmission can be removed from the engine, if necessary.

Transmission Jacks

Transmission jacks are designed to help you while removing a transmission from under the vehicle. The weight of the transmission makes it difficult and unsafe to remove it without much assistance or a transmission jack. These jacks fit under the transmission (Figure 3) and are typically equipped with hold-down chains. These chains are used to secure the transmission to

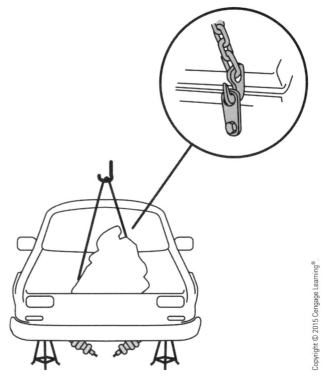

Figure 2 A lifting chain attached to the engine's eye plates.

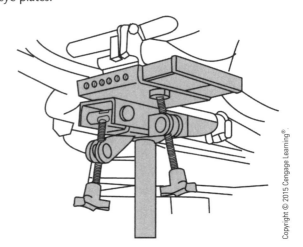

Figure 3 A typical transmission jack in place under a transmission.

the jack. The transmission's weight rests on the jack's saddle.

Transmission jacks are available in two basic styles. One is used when the vehicle is raised by a hydraulic jack and is set on jack stands. The other style is used when the vehicle is raised on a lift.

Transaxle Removal and Installation Equipment

Removal and replacement of transversely mounted engines may require special tools. The engines of

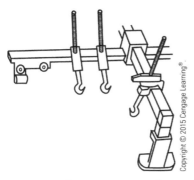

Figure 4 A typical engine support fixture for front-wheel-drive vehicles.

some FWD vehicles are removed by lifting them from the top. Others must be removed from the bottom, which requires different equipment. Make sure you follow the instructions given by the manufacturer and use the appropriate tools and equipment. The required equipment varies with manufacturer and vehicle model; however, most tools accomplish the same thing.

To remove the engine and transmission from under the vehicle, the vehicle must be raised. A crane and/or support fixture (Figure 4) is used to hold the engine and transaxle assembly in place while the assembly is being readied for removal. When everything is set for removal of the assembly, the crane is used to lower the assembly onto a cradle. The cradle is similar to a hydraulic floor jack and is used to lower the assembly further so it can be rolled out from under the vehicle. The transaxle can be separated from the engine once it has been removed from the vehicle.

When the transaxle is removed as a single unit, the engine must be supported while it is in the vehicle before, during, and after transaxle removal. Special fixtures mount to the vehicle's upper frame or suspension parts. These supports have a bracket that is attached to the engine. With the bracket in place, the engine's weight is now on the support fixture, and the transmission can be removed.

Transmission or Transaxle Holding Fixtures

Special holding fixtures should be used to support the transmission or transaxle after it has been removed from the vehicle. These holding fixtures, which may be standalone units or bench mounted, allow the transmission to be repositioned easily during repair work.

Machinist's Rule

A machinist's rule is very much like an ordinary ruler. Each edge of this measuring tool is divided into increments based on a different scale. A typical machinist's rule based on the United States Customary System (USCS) system of measurement may have scales based on 1/8-, 1/16-, 1/32-, and 1/64-inch intervals. Of course, metric machinist rules are also available. Metric rules are usually divided into 0.5-mm and 1-mm increments.

Some machinist's rules are based on decimal intervals. These are typically divided into 1/10-, 1/50-, and 1/1,000-inch (0.1, 0.01, and 0.001) increments. When measuring dimensions that are specified in decimals, decimal machinist's rules are very helpful because you won't need to convert fractions to decimals.

Micrometers

A micrometer is used to measure linear outside and inside dimensions. Both outside and inside micrometers are calibrated and read in the same manner. The major components and markings of a micrometer include the frame, anvil, spindle, lock nut, sleeve, sleeve numbers, sleeve long line, thimble marks, thimble, and ratchet. Micrometers are calibrated in either inch or metric graduations and are available in a range of sizes.

To use and read a micrometer, choose the appropriate size for the object being measured. Typically they measure an inch, therefore the range covered by one-size micrometer measures from 0 to 1 inch, another measures 1 to 2 inches, and so on.

Open the jaws of the micrometer and slip the object between the spindle and the anvil (Figure 5). While holding the object against the

Figure 5 A micrometer is used to measure shim thickness.

anvil, turn the thimble using your thumb and forefinger until the spindle contacts the object. Never clamp the micrometer tightly—use only enough pressure on the thimble to allow the work to just fit between the anvil and spindle. To get accurate readings, you should slip the micrometer back and forth over the object until you feel a very light resistance, while at the same time rocking the tool from side to side to make certain the spindle cannot be closed any further. When a satisfactory adjustment has been made, lock the micrometer. Read the measurement scale.

The graduations on the sleeve each represent 0.025 inch. To read a measurement on a micrometer, begin by counting the visible lines on the sleeve and multiplying them by 0.025. The graduations on the thimble assembly define the area between the lines on the sleeve. The number indicated on the thimble is added to the measurement shown on the sleeve; the sum is the dimension of the object.

Micrometers are available to measure in 0.0001 (ten-thousandths) of an inch. Use this type of micrometer if the specifications call for this much accuracy.

A metric micrometer is read in the same way, except that the graduations are expressed in the metric system of measurement. Each number on the sleeve represents 5 millimeters (mm), or 0.005 meter (m). Each of the 10 equal spaces between each number, with index lines alternating above and below the horizontal line, represents 0.5 mm, or five-tenths of an mm. Therefore, one revolution of the thimble changes the reading one space on the sleeve scale or 0.5 mm. The beveled edge of the thimble is divided into 50 equal divisions with every fifth line numbered: 0, 5, 10, up to 45. Since one complete revolution of the thimble advances the spindle 0.5 mm, each graduation on the thimble is equal to one hundredth of a millimeter. As with the inch-graduated micrometer, the separate readings are added together to obtain the total reading.

Some technicians use a digital micrometer, which is easier to read. These tools do not have the various scales; instead, the measurement is displayed and read directly off the micrometer.

Inside micrometers can be used to measure the inside diameter of a bore. To do this, place the tool inside the bore and extend the measuring surfaces until each end touches the bore's surface. If the bore is large, it might be necessary to use an extension rod to increase the micrometer's range. These extension rods come in various lengths. An inside micrometer is read in the same manner as an outside micrometer.

A depth micrometer is used to measure the distance between two parallel surfaces. The sleeves, thimbles, and ratchet screws operate as they do in other micrometers. Depth micrometers are read in the same way as other micrometers.

If a depth micrometer is used with a gauge bar, it is important to keep both the bar and the micrometer from rocking. Any movement of either part results in an inaccurate measurement.

Telescoping Gauge

Telescoping gauges are used for measuring bore diameters and other clearances. They may also be called snap gauges. Telescoping gauges are available in sizes ranging from fractions of an inch through 6 inches. Each gauge consists of two telescoping plungers, a handle, and a lock screw. Snap gauges are normally used with an outside micrometer.

To use a telescoping gauge, insert it into the bore and loosen the lock screw. This will allow the plungers to snap against the bore. Once the plungers have expanded, tighten the lock screw. Then remove the gauge and measure the expanse with a micrometer.

Small Hole Gauge

A small hole or ball gauge works just like a telescoping gauge, but it is designed for small bores. After it is placed into the bore and expanded, it is removed and measured with a micrometer. Like the telescoping gauge, the small hole gauge consists of a lock, a handle, and an expanding end. The end is made to expand or retract by turning the gauge handle.

Feeler Gauge

A feeler gauge is a thin strip of metal or plastic of known and closely controlled thickness. Several of these strips are often assembled together as a feeler gauge set that looks like a pocketknife. The desired thickness gauge can be pivoted away from the others for convenient use. A feeler gauge set usually contains strips or leaves of 0.002- to 0.010-inch thickness (in steps of 0.001 inch) and leaves of 0.012- to 0.024-inch thickness (in steps of 0.002 inch).

A feeler gauge can be used by itself to measure oil pump clearances, gear clearances, endplay, and other distances. It can also be used with a precision straightedge to check the flatness of a sealing surface.

Straightedge

A straightedge is no more than a flat bar machined to be totally flat and straight. To be effective, it must be flat and straight. Any surface that should be flat can be checked with a straightedge and feeler gauge set. The straightedge is placed across and at angles on the surface. At any low points on the surface, a feeler gauge can be placed between the straightedge and the surface. The size gauge that fills in the gap is the amount of warpage or distortion.

Dial Indicator

The dial indicator is calibrated in 0.001-inch (one-thousandth-inch) increments. Metric dial indicators are also available. Both types are used to measure movement. Common uses of the dial indicator include measuring backlash (Figure 6), endplay (Figure 7), and flywheel and axle flange runout.

To use a dial indicator, position the indicator rod against the object to be measured. Push the indicator toward the work until the indicator needle travels far enough around the gauge face to permit movement to be read in either direction. Zero the indicator needle on the gauge. Move the object in the direction required while observing the needle of the gauge. Always be sure the range of the dial indicator is sufficient to allow the amount of movement required by the measuring procedure. For example, never use a 1-inch indicator on a component that can move 2 inches.

Torque Wrench

Torque is the twisting force used to turn a fastener against the friction between the threads and

Figure 6 Checking backlash with a dial indicator.

Figure 7 Checking axle shaft endplay with a dial indicator.

between the head of the fastener and the surface of the component. The fact that practically every vehicle and engine manufacturer publishes a list of torque recommendations is ample proof of the importance of using proper amounts of torque when tightening nuts or bolts. The amount of torque applied to a fastener is measured with a torque-indicating wrench or torque wrench.

A torque wrench is basically a ratchet or breaker bar with some means of displaying the amount of torque exerted on a bolt when pressure is applied to the handle. Torque wrenches are available with the various drive sizes. Sockets are inserted onto the drive and then placed over the bolt. As pressure is exerted on the bolt, the torque wrench indicates the amount of torque.

The common types of torque wrenches are available with inch-pound and foot-pound increments.

■ A beam torque wrench is not highly accurate. It relies on a beam metal that points to the torque reading.

■ A "click"-type torque wrench clicks when the desired torque is reached. The handle is twisted to set the desired torque reading.

■ A dial torque wrench has a dial that indicates the torque exerted on the wrench. The wrench may have a light or buzzer that turns on when the desired torque is reached.

A digital readout type displays the torque and is commonly used to measure turning effort, as well as for tightening bolts. Some designs of this type torque wrench have a light or buzzer that turns on when the desired torque is reached.

Blowgun

Blowguns are used for cleaning parts. Never point a blowgun at yourself or someone else. A blowgun snaps into one end of an air hose and directs airflow when a button is pressed. Always use an OSHA-approved air blowgun. Before using a blowgun, be sure it has not been modified to eliminate air-bleed holes on the side.

Clutch Alignment Tool

To keep the clutch disc centered on the flywheel while assembling the clutch, a clutch alignment

Figure 8 A clutch alignment tool in place.

Copyright © 2015 Cengage Learning®

tool is used (Figure 8). The tool is inserted through the input shaft opening of the pressure plate and is passed through the clutch disc. The tool is then inserted into the pilot bushing or bearing. The outer diameter (OD) of the alignment tool that goes into the pilot must be only slightly smaller than the inner diameter (ID) of the pilot bushing. The OD of the tool that holds the disc in place also must be only slightly smaller than the ID of the disc's splined bore. The effectiveness of this tool depends on its diameters, which is why it is best to have various sizes of clutch alignment tools.

Gear and Bearing Pullers

Many tools are designed for a specific purpose. An example of a special tool is a gear and bearing puller. Many gears and bearings have a slight interference fit (press fit) when they are installed on a shaft (Figure 9) or in a housing. Something that has a press fit has an interference fit; for example, the ID of a bore is 0.001 inch smaller than the outside diameter of a shaft, so when the shaft is fitted into the bore it must be pressed in to overcome the 0.001-inch interference. This press fit prevents the parts from moving against each other. These gears and bearings must be removed carefully to prevent damage to the gears, bearings, or shafts. Prying or hammering can break or bind the parts. A puller with the proper jaws and adapters should be used to remove gears and bearings. With the proper puller, the force required to remove a gear or bearing can be applied with a slight and steady motion.

Figure 9 Removing side bearings with a threaded puller.

Bushing and Seal Pullers and Drivers

Another commonly used group of special tools are the various designs of bushing and seal drivers and pullers. Pullers are either a threaded or slide hammer type of tool. Always make sure you use the correct tool for the job because bushings and seals are easily damaged if the wrong tool or procedure is used. Car manufacturers and specialty tool companies work together closely to design and manufacture special tools for repairing cars. Most of these special tools are listed in the appropriate service manuals.

A commonly used bearing puller is a clutch pilot bearing and bushing remover. Typically, the tool is designed to hook into the back of the bearing. Once it is in place, the threaded portion is tightened and the bearing pulled out. The new pilot bearing or bushing is driven into the bore with a hammer and correctly sized driver.

Presses

Many transmission and driveline repairs require the use of a powerful force to assemble or disassemble parts that are press fit together. Removal and installation of axle and final drive bearings (Figure 10), universal joint replacement, and transmission assembly work are just a few of the examples. Presses can be hydraulic, electric, air, or hand driven. Capacities range up to 150 tons of pressing force, depending on the size and design of the press. Smaller arbor and C-frame presses can be bench or pedestal mounted, whereas high-capacity units are freestanding or floor mounted.

Figure 10 Installing bearings onto a differential case with a driver and hydraulic press.

Universal Joint Tools

Although universal joints can be serviced with hand tools and a vise, many technicians prefer to use specifically designed tools. One such tool is a C-clamp modified to include a bore that allows the joint's caps to slide in while the clamp is tightened over an assembled joint to remove it. Other tools are the various drivers used with a press to press the joint in and out of its yoke.

Retaining Ring Pliers

Often, a transmission technician will run into many different styles and sizes of retaining rings that hold subassemblies together or keep them in a fixed location. Using the correct tool to remove and install these rings is the only safe way to work with them. All transmission and driveline technicians should have an assortment of retaining ring pliers.

Special Tool Sets

Vehicle manufacturers and specialty tool companies work closely together to design and manufacture special tools required to repair transmissions and final drives. Most of these special tools are listed in the appropriate service manuals and are part of each manufacturer's essential tool kit.

Service Information

Perhaps the most important tool you will use is the appropriate service information. There is no way a technician can remember all the procedures and specifications needed to repair all vehicles. Thus, a good technician relies on service information sources. Good information plus knowledge allows a technician to fix a problem with the least frustration and at the lowest expense to the customer.

To obtain the correct transmission specifications and other information, you must first identify the transmission you are working on. The best source for positive identification is the vehicle identification number (VIN). The transmission code can be interpreted through information given in the service information. The information may also help you identify the transmission through appearance, casting numbers, and markings on the housing.

The primary source of repair and specification information for any car, van, or truck is the manufacturer. The manufacturer releases service information each year for every vehicle it builds. Because of the enormous amount of information, some manufacturers release service information more than once per year per car model. The information is typically divided into sections based on the major systems of the vehicle. In the case of transmissions, there is a section for each transmission that may be in the vehicle. The service information from the manufacturers covers all repairs, adjustments, specifications, detailed diagnostic procedures, and special tools.

Since many technical changes occur on specific vehicles each year, the service information needs to be constantly updated. Updates are published as service bulletins (often referred to as technical service bulletins, or TSBs) that show the changes in specifications and repair procedures during the model year. The car manufacturer provides these bulletins to dealers and repair facilities on a regular basis.

Service information is also made available by independent companies, rather than the manufacturers, but they pay for and get most of their information from the car makers. Information from these companies is usually condensed and is more general in nature than the information from the manufacturer. The condensed format allows for more coverage in less space and, therefore, is not always specific. Many of the larger parts manufacturers have excellent guides on the various parts they manufacture or supply. They also provide updated service bulletins on their products. Other sources for up-to-date technical information are trade magazines and trade associations.

Service information is now commonly found electronically on DVDs and the Internet. Using electronics to find information is easier and quicker than using a manual. The information on the Internet is updated as soon as new information is released and it not only contains the most recent service bulletins but also engineering and field service fixes. All a technician needs to do is enter vehicle information and then move to the appropriate part or system. Once the information is retrieved, a technician can read it off the screen or print it out and take it to the service bay.

JOB SHEETS
REQUIRED SUPPLEMENTAL TASKS (RST)

JOB SHEET 1

Shop Safety Survey

Name _____ Station _____ Date _____

NATEF Correlation

This Job Sheet addresses the following **RST** tasks:

Shop and Personal Safety:

1. Identify general shop safety rules and procedures.

8. Identify the location and use of eyewash stations.

10. Comply with the required use of safety glasses, ear protection, gloves, and shoes during lab/shop activities

11. Identify and wear appropriate clothing for lab/shop activities

12. Secure hair and jewelry for lab/shop activities.

Objective

As a professional technician, safety should be one of your first concerns. This job sheet will increase your awareness of shop safety rules and safety equipment. As you survey your shop area and answer the following questions, you will learn how to evaluate the safeness of your workplace.

Materials

Copy of the shop rules from the instructor

PROCEDURE

Your instructor will review your progress throughout this worksheet and should sign off on the sheet when you complete it.

1. Have your instructor provide you with a copy of the shop rules and procedures.

   Have you read and understood the shop rules? ☐ Yes ☐ No

2. Before you begin to evaluate your work area, evaluate yourself. Are you dressed to work safely? ☐ Yes ☐ No

   If no, what is wrong? _____

3. Are your safety glasses OSHA approved? ☐ Yes ☐ No

   Do they have side protection shields? ☐ Yes ☐ No

4. Look around the shop and note any area that poses a potential safety hazard.

   _____

   _____

   All true hazards should be brought to the attention of the instructor immediately.

5. What is the air line pressure in the shop? _____ psi

   What should it be? _____ psi

6. Where is the first-aid kit(s) kept in the work area?

   _____

   _____

7. Ask the instructor to show the location of, and demonstrate the use of, the eyewash station. Where is it and when should it be used?

   _____

   _____

8. What is the shop's procedure for dealing with an accident?

   _____

   _____

   _____

9. Explain how to secure hair and jewelry while working in the shop.

   _____

10. List the phone numbers that should be called in case of an emergency.

    _____

Problems Encountered

_____

_____

_____

Instructor's Comments

_____

_____

_____

JOB SHEET 2

Working in a Safe Shop Environment

Name _____ Station _____ Date _____

NATEF Correlation

This Job Sheet addresses the following **RST** tasks:

Shop and Personal Safety:

2. Utilize safe procedures for handling of tools and equipment.

3. Identify and use proper placement of floor jacks and jack stands.

4. Identify and use proper procedures for safe lift operation.

5. Utilize proper ventilation procedures for working within the lab/shop area.

6. Identify marked safety areas.

9. Identify the location of the posted evacuation routes.

Objective

This job sheet will help you work safely in the shop. Two of the basic tools a technician uses are lifts and the floor jack.

This job sheet also covers the technicians' environment.

Materials

Vehicle for hoist and jack stand demonstration
Service information

Describe the vehicle being worked on:

Year _____ Make _____ Model _____

VIN _____ Engine type and size _____

PROCEDURE

1. Are there safety areas marked around grinders and other machinery? ☐ Yes ☐ No

2. Are the shop emergency escape routes clearly marked? ☐ Yes ☐ No

3. Have your instructor demonstrate the exhaust gas ventilation system in the shop. Explain the importance of the ventilation system.

_____

_____

4. What types of Lifts are used in the shop?

   _____

   _____

5. Find the location of the correct lifting points for the vehicle supplied by the instructor. On the rear of this sheet, draw a simple figure showing where these lift points are.

6. Ask your instructor to demonstrate the proper use of the lift.

   Summarize the proper use of the lift.

   _____

   _____

   _____

   _____

7. Demonstrate the proper use of jack stands with the help of your instructor.

   Summarize the proper use of jack stands.

   _____

   _____

   _____

Problems Encountered

_____

_____

_____

Instructor's Comments

_____

_____

_____

JOB SHEET 3

Fire Extinguisher Care and Use

Name _____ Station _____ Date _____

NATEF Correlation

This Job Sheet addresses the following **RST** task:

Shop and Personal Safety:

7. Identify the location and the types of fire extinguishers and other fire safety equipment; demonstrate knowledge of the procedures for using fire extinguishers and other fire safety equipment.

Objective

Upon completion of this job sheet, you will be able to demonstrate knowledge of the procedures for using fire extinguishers and other fire safety equipment, and identify the location of fire extinguishers in the shop.

> **NOTE:** *Never fight a fire that is out of control or too large. Call the fire department immediately!*

1. Identify the location of the fire extinguishers in the shop.

   _____

   _____

2. Have the fire extinguishers been inspected recently? (Look for a dated tag.)

   _____

   _____

3. What types of fires are each of the shop's fire extinguishers rated to fight?

   _____

   _____

4. What types of fires should not be used with the shop's extinguishers?

   _____

   _____

5. One way to remember the operation of a fire extinguisher is to remember the term PASS. Describe the meaning of PASS on the following lines.

   a. P

   _____

   b. A

   _____

c. S

_____

d. S

_____

Problems Encountered

_____

_____

_____

Instructor's Comments

_____

_____

_____

JOB SHEET 4

Working Safely Around Air Bags

Name _____ Station _____ Date _____

NATEF Correlation

This Job Sheet addresses the following **RST** task:

Shop and Personal Safety:

13. Demonstrate awareness of the safety aspects of supplemental restraint systems (SRS), electronic brake control systems, and hybrid vehicle high-voltage circuits.

Objective

Upon completion of this job sheet, you should be able to work safely around and with air bag systems.

Tools and Materials

A vehicle(s) with air bag

Safety glasses, goggles

Service information appropriate to vehicle(s) used

Describe the vehicle being worked on:

Year _____ Make _____ Model _____

VIN _____ Engine type and size _____

PROCEDURE

1. Locate the information about the air bag system in the service information. How are the critical parts of the system identified in the vehicle?

   _____

   _____

2. List the main components of the air bag system and describe their location.

   _____

   _____

3. There are some very important guidelines to follow when working with and around air bag systems. Look through the service information to find the answers to the questions and fill in the blanks with the correct words.

   a. Wait at least _____ minutes after disconnecting the battery before beginning any service. The reserve _____ module is capable of storing enough energy to deploy the air bag for up to _____ minutes after battery voltage is lost.

b. Never carry an air bag module by its _____ or _____, and, when carrying it, always face the trim and air bag _____ from your body. When placing a module on a bench, always face the trim and air bag _____.

c. Deployed air bags may have a powdery residue on them. _____ is produced by the deployment reaction and is converted to _____ when it comes in contact with the moisture in the atmosphere. Although it is unlikely that harmful chemicals will still be on the bag, it is wise to wear _____ and _____ when handling a deployed air bag. Immediately wash your hands after handling a deployed air bag.

d. A live air bag must be _____ before it is disposed. A deployed air bag should be disposed of in a manner consistent with the _____ and manufacturer's procedures.

e. Never use a battery- or AC-powered _____, _____, or any other type of test equipment in the system unless the manufacturer specifically says to. Never probe with a _____ for voltage.

4. Explain how an air bag sensor should be handled before it is installed on the vehicle.

Problems Encountered

_____

_____

_____

Instructor's Response

_____

_____

_____

JOB SHEET 5

High-Voltage Hazards in Today's Vehicles

Name _____ Station _____ Date _____

NATEF Correlation

This Job Sheet addresses the following **RST** task:

Shop and Personal Safety:

14. Demonstrate awareness of the safety aspects of high-voltage circuits (such as high intensity discharge (HID) lamps, ignition systems, injection systems, etc.).

Objective

Upon completion of this job sheet, you will be able to describe some of the necessary precautions to take in performing work around high-voltage hazards such as HID headlamps and ignition systems.

HID Headlamp Precautions

Describe the vehicle being worked on:

Year _____ Make _____ Model _____

VIN _____ Engine type and size _____

Materials needed

Service information for HID headlamps

Describe general operating condition:

1. List three precautions a technician should observe when working with HID headlamp systems.

   _____

   _____

   _____

2. How many volts are necessary to initiate and maintain the arc inside the bulb of this HID system?

   _____

   _____

3. A technician must never probe with a test lamp between the HID ballast and the bulb. Explain why?

   _____

High-Voltage Ignition System Precautions

4. High-voltage ignition systems can cause serious injury, especially for those who have heart problems. Name at least three precautions to take when working around high-voltage ignition systems.

_____

_____

_____

_____

Problems Encountered

_____

_____

_____

Instructor's Comments

_____

_____

_____

JOB SHEET 6

Hybrid High-Voltage and Brake System Pressure Hazards

Name _____ Station _____ Date _____

NATEF Correlation

This Job Sheet addresses the following **RST** task:

Shop and Personal Safety:

13. Demonstrate awareness of the safety aspects of supplemental restraint systems (SRS), electronic brake control systems, and hybrid vehicle high-voltage circuits.

Objective

Upon completion of this job sheet, you will be able to describe some of the necessary precautions to take while performing work around high-voltage systems such as that used in hybrid vehicles, as well as the high-pressure systems of some braking control systems.

Tools and Materials

Appropriate service information

AUTHOR'S NOTE: *According to the vehicles' manufacturers, a technician should not work on hybrid vehicles without having the specific training, which is beyond the scope of this job sheet. This job sheet assumes that the student will be accessing information only, not actually working on live high-voltage vehicles.*

Protective Gear

Goggles or safety glasses with side shields

High-voltage gloves with properly inspected liners

Orange traffic cones to warn others in the shop of a high-voltage hazard

Describe the vehicle being worked on:

Year _____ Make _____ Model _____

VIN _____ Engine type and size _____

Describe general operating condition:

_____

PROCEDURE

High Pressure Braking Systems:

NOTE: *Many vehicles have a high pressure accumulator or a high pressure pump in their braking systems. Opening one of these systems can be hazardous due to the high pressures involved.*

1. Research the vehicle you have been assigned and describe the procedure that must be followed BEFORE opening the hydraulic braking system.

   _____

   _____

   _____

Hybrid Vehicles:

1. Special gloves with liners are to be inspected before each use when working on a hybrid vehicle.

   A. How should the integrity of the gloves be checked?

   _____

   _____

   B. How often should the gloves be tested (at a lab) for recertification?

   _____

   C. What precautions must be taken when storing the gloves?

   _____

   _____

   _____

   D. When must the gloves be worn?

   _____

   _____

   _____

2. What color are the high-voltage cables on hybrid vehicles?

   _____

3. What must be done BEFORE disconnecting the main voltage supply cable?

   _____

   _____

   _____

4. Describe the safety basis of the "one hand rule."

_____

_____

_____

5. Explain the procedure to disable high voltage on the vehicle you selected.

_____

_____

_____

6. Explain the procedure to test for high voltage to ensure that the main voltage is disconnected.

_____

_____

Problems Encountered

_____

_____

_____

Instructor's Comments

_____

_____

_____

JOB SHEET 7

Material Data Safety Sheet Usage

Name _____ Station _____ Date _____

NATEF Correlation

This Job Sheet addresses the following **RST** task:

Shop and Personal Safety:

12. Locate and demonstrate knowledge of material safety data sheets (MSDS).

Objective

On completion of this job sheet, the student will be able to locate the MSDS folder and describe the use of an MSDS sheet on the job site.

Materials

Selection of chemicals from the shop

MSDS sheets

1. Locate the MSDS folder in the shop. It should be in a prominent location. Did you have any problems finding the folder?

   _____

   _____

2. Pick a common chemical from your tool room such as brake cleaner. Locate the chemical in the MSDS folder. What chemical did you choose?

   _____

   _____

3. What is the flash point of the chemical? _____

4. Briefly describe why the flash point is important.

   _____

   _____

5. What is the first aid if the chemical is ingested?

   _____

   _____

6. Can this chemical be absorbed through the skin?

   _____

   _____

7. What are the signs of exposure to the chemical you selected?

_____

_____

8. What are the primary health hazards of the chemical?

_____

_____

9. What is the first aid procedure for exposure to this chemical?

_____

_____

10. What are the recommendations for protective clothing?

_____

_____

Problems Encountered

_____

_____

_____

Instructor's Comments

_____

_____

_____

JOB SHEET 8

Measuring Tools and Equipment Use

Name _____ Station _____ Date _____

NATEF Correlation

This Job Sheet addresses the following **RST** tasks:

Tools and Equipment:

1. Identify tools and their usage in automotive applications.

2. Identify standard and metric designations.

3. Demonstrate safe handling and use of appropriate tools.

4. Demonstrate proper cleaning, storage, and maintenance of tools and equipment.

5. Demonstrate proper use of precision measuring tools (i.e., micrometer, dial-indicator, dial caliper).

Objective

Upon completion of this job sheet, you will be able to make measurements using micrometers, dial indicators, pressure gauges, and other measuring tools. You will also be able to demonstrate the safe handling and use of appropriate tools, and demonstrate their proper use.

Tools and Materials

Items to measure selected by the instructor

Precision measuring tools: micrometer (digital or manual), dial caliper, vacuum or pressure gauge (selected by the instructor)

PROCEDURE

Have your instructor demonstrate the measuring tools that are available to you in your shop. The tools should include both standard and metric tools.

1. Describe the measuring tools you will be using.

   _____

   _____

2. Describe the items that you will be measuring.

   _____

   _____

3. Describe any special handling or safety procedures that are necessary when using the tools.

   _____

   _____

4. Describe any special cleaning or storage that these tools might require.

   _____

   _____

5. Describe the metric unit of measurement of the tools, such as millimeters, centimeters, kilopascals, etc.

   _____

   _____

6. Describe the standard unit of measure of the tools, such as inches, pounds per square inch, etc.

   _____

   _____

7. Measure the components, list them, and record your measurements in the following table.

Item measured	Measurement taken	Unit of measure

8. Clean and store the tools. Describe the process next.

Problems Encountered

_____

_____

_____

Instructor's Comments

_____

_____

_____

JOB SHEET 9

Preparing the Vehicle for Service and Customer

Name _____ Station _____ Date _____

NATEF Correlation

This Job Sheet addresses the following **RST** tasks:

Preparing Vehicle for Service:

1. Identify information needed and the service requested on a repair order.

2. Identify purpose and demonstrate proper use of fender covers, mats.

3. Demonstrate use of the three C's (concern, cause, and correction).

4. Review vehicle service history.

5. Complete work order to include customer information, vehicle identifying information, customer concern, related service history, cause, and correction.

Preparing Vehicle for Customer:

1. Ensure vehicle is prepared to return to customer per school/company policy (floor mats, steering wheel cover, etc.).

Objective

Upon completion of this job sheet, you will be able to prepare a service work order based on customer input, vehicle information, and service history. The student will also be able to describe the appropriate steps to take to protect the vehicle and delivering the vehicle to the customer after the repair.

Tools and Materials

An assigned vehicle or the vehicle of your choice

Service work order or computer-based shop management package

Parts and labor guide

Work Order Source: Describe the system used to complete the work order. If a paper repair order is being used, describe the source.

_____

_____

PROCEDURE

1. Prepare the shop management software for entering a new work order or obtain a blank paper work order. Describe the type of repair order you are going to use.

   _____

   _____

2. Enter customer information, including name, address, and phone numbers onto the work order.   Task Completed ☐

3. Locate and record the vehicle's VIN. Where did you find the VIN?

   _____

4. Enter the necessary vehicle information, including year, make, model, engine type and size, transmission type, license number, and odometer reading.   Task Completed ☐

5. Does the VIN verify that the information about the vehicle is correct?

   _____

6. Normally, you would interview the customer to identify his or her concerns. However, to complete this job sheet, assume the only concern is that the customer wishes to have the front brake pads replaced. Also, assume no additional work is required to do this. Add this service to the work order.   Task Completed ☐

7. Prepare the vehicle for entering the service department. Add floor mats, seat covers, and steering wheel covers to the vehicle.   Task Completed ☐

8. The history of service to the vehicle can often help diagnose problems, as well as indicate possible premature part failure. Gathering this information from the customer can provide some of the data needed. For this job sheet, assume the vehicle has not had a similar problem and was not recently involved in a collision. Service history is further obtained by searching files for previous service. Often this search is done by customer name, VIN, and license number. Check the files for any related service work.   Task Completed ☐

9. Search for technical service bulletins on this vehicle that may relate to the customer's concern. Did you find any? _____ If so, record the reference numbers here.

   _____

   _____

10. Based on the customer's concern, service history, TSBs, and your knowledge, what is the likely cause of this concern?

    _____

    _____

11. Add this information to the work order.   Task Completed ☐

12. Prepare to make a repair cost estimate for the customer. Identify all parts that may need to be replaced to correct the concern. List these here.

_____

_____

13. Describe the task(s) that will be necessary to replace the part.

_____

_____

14. Using the parts and labor guide, locate the cost of the parts that will be replaced and enter the cost of each item onto the work order at the appropriate place for creating an estimate. If the valve or cam cover is leaking, what part will need to be replaced?

_____

15. Now, locate the flat rate time for work required to correct the concern. List each task with its flat rate time.

_____

_____

16. Multiply the time for each task by the shop's hourly rate and add the cost of each item to the work order at the appropriate place for creating an estimate. Ask your instructor which shop labor rate to use and record it here.

_____

17. Many shops have a standard amount they charge each customer for shop supplies and waste disposal. For this job sheet, use an amount of ten dollars for shop supplies.   Task Completed ☐

18. Add the total costs and insert the sum as the subtotal of the estimate.   Task Completed ☐

19. Taxes must be included in the estimate. What is the sales tax rate and does it apply to both parts and labor, or just one of these?

_____

_____

20. Enter the appropriate amount of taxes to the estimate, then add this to the subtotal. The end result is the estimate to give the customer.   Task Completed ☐

21. By law, how accurate must your estimate be?

_____

_____

22. Generally speaking, the work order is complete and is ready for the customer's signature. However, some businesses require additional information; make sure you add that information to the work order. On the work order, there is a legal statement that defines what the customer is agreeing to. Briefly describe the contents of that statement.

_____

23. Now that the vehicle is completed and it is ready to be returned to the customer, what are the appropriate steps to take to deliver the vehicle to the customer? What would you do to make the delivery special? What should not happen when the vehicle is delivered to the customer?

_____

_____

Problems Encountered

_____

_____

_____

Instructor's Comments

_____

_____

_____

JOB SHEETS

MANUAL DRIVE TRAIN AND AXLES JOB SHEET 10

Identifying Problems and Concerns

Name _____ Station _____ Date _____

NATEF Correlation

This Job Sheet addresses the following **AST/MAST** task:

A.1. Identify and interpret drive train concerns; determine necessary action.

Objective

Upon completion of this job sheet, you will be able to identify and interpret drive train concerns, prior to diagnosing or testing the systems.

Tools and Materials

Appropriate service information

Protective Clothing

Goggles or safety glasses with side shields

Describe the vehicle being worked on:

Year _____ Make _____ Model _____

VIN _____ Engine type and size _____

PROCEDURE

1. Start the engine and describe how the engine seems to be running:

   _____

   _____

2. Take the vehicle for a safe road test and pay strict attention to the noises and vibrations during all conditions (acceleration, deceleration, cruising, at idle, through turns, and on different road surfaces). Describe your results here.

_____

_____

_____

_____

3. If noises or vibrations were felt and/or heard, state what you think the cause may be based on, and the occurrence and apparent location of the noise or vibration.

_____

_____

_____

_____

4. If you experienced any abnormal noise or vibration during the road test, carefully inspect the vehicle's tires, suspension and steering system, and exhaust system for evidence that they are causing the problem. Describe the results of your inspection here.

_____

_____

_____

_____

5. On the road test, did you experience any difficulty changing gears? If so, what happened?

_____

_____

_____

_____

6. Did you experience a noise or other problem only while you were traveling in one particular gear? If so, what gear? What happened?

_____

_____

_____

_____

7. Based on your inspection and road test, what can you conclude about this vehicle's clutch assembly?

_____

_____

_____

_____

8. Based on your inspection and road test, what can you conclude about this vehicle's transmission/transaxle?

_____

_____

_____

_____

9. Based on your inspection and road test, what can you conclude about this vehicle's final drive unit?

_____

_____

_____

_____

10. Based on your inspection and road test, what can you conclude about this vehicle's drive axles and drive train?

_____

_____

_____

_____

Problems Encountered

_____

_____

_____

Instructor's Comments

_____

_____

_____

MANUAL DRIVE TRAIN AND AXLES JOB SHEET 11

Gathering Vehicle Information

Name _____ Station _____ Date _____

NATEF Correlation

This Job Sheet addresses the following **MLR** task:

A.1. Research applicable vehicle and service information, such as drive train system operation, fluid type, vehicle service history, service precautions, and technical service bulletins.

This Job Sheet addresses the following **AST/MAST** task:

A.2. Research applicable vehicle and service information, such as drive train system operation, fluid type, vehicle service history, service precautions, and technical service bulletins.

Objective

Upon completion of this job sheet, you will be able to gather service information about a vehicle and its drive train and manual transmission or transaxle.

Tools and Materials

Appropriate service information

Computer

Protective Clothing

Goggles or safety glasses with side shields

Describe the vehicle being worked on:

Year _____ Make _____ Model _____

VIN _____

PROCEDURE

1. Visually inspect the vehicle and describe the major components and the location of the drive train.

   _____

   _____

   _____

   _____

2. Briefly describe the operation of the transmission in your assigned vehicle.

   _____

   _____

   _____

   _____

3. While looking in the engine compartment or under the vehicle, locate the identification tag on the transmission or transaxle. Describe where you found it.

   _____

   _____

4. Summarize the information contained on this label.

   _____

   _____

   _____

5. If the final drive housing is separate from the transmission or transaxle housing, locate the identification tag on the final drive unit. Describe where you found it.

   _____

   _____

6. Summarize the information found on the final drive identification tag.

   _____

   _____

   _____

7. Using the service information, locate the information about the vehicle's drive train. Describe the clutch operation and any controls and/or sensors attached directly to parts of the drive train.

   _____

   _____

   _____

   _____

8. Using the service information, locate and record all service precautions regarding the drive train noted by the manufacturer.

   _____

   _____

   _____

9. Refer to the service information and identify the exact type of fluid that should be used in this transmission and final drive unit. What type is it?

_____

10. Using the available information, locate and record the vehicle's service history. If service history for the vehicle is not available, as the owner about any recent repairs to the vehicle.

_____

_____

_____

_____

11. Using the available information sources, summarize all Technical Service Bulletins for this vehicle that relate to the transmission and drive train.

_____

_____

_____

_____

Problems Encountered

_____

_____

_____

Instructor's Comments

_____

_____

_____

MANUAL DRIVE TRAIN AND AXLES JOB SHEET 12

Drive Train Fluid Service

Name _____ Station _____ Date _____

NATEF Correlation

This Job Sheet addresses the following **MLR** tasks:

A.2. Drain and fill manual transmission/transaxle and final drive unit.

A.3. Check fluid condition; check for leaks.

This Job Sheet addresses the following **AST/MAST** tasks:

A.3. Diagnose fluid loss, level, and condition concerns; determine necessary action.

A.4. Drain and fill manual transmission/transaxle and final drive unit.

Objective

Upon completion of this job sheet, you will be able to diagnose the cause of excessive fluid loss and contaminated fluid. You will also be able to drain and fill a manual transmission/transaxle and final drive unit with the correct fluid.

Tools and Materials

Appropriate service information   Drain pan
Lift   Clean rags

Protective Clothing

Goggles or safety glasses with side shields

Describe the vehicle being worked on:

Year _____ Make _____ Model _____

VIN _____

PROCEDURE

1. Raise the vehicle and visually inspect the vehicle's drive train for signs of fluid leakage. Summarize your results.

   _____

   _____

   _____

   _____

2. Based on the preceding step, what service do you recommend?

   _____

   _____

3. If fluid is leaking from around the seal at the transmission's extension housing, what should you check before replacing the seal?

_____

_____

4. List at least five possible sources for a fluid leak in the drive train of this vehicle.

_____

_____

_____

_____

_____

5. Explain why a fluid leak can also cause the fluid to become contaminated with water or dirt.

_____

_____

_____

_____

6. Locate the preventive maintenance schedule for the drive train in the service information or the vehicle's owner's manual. Summarize what should be done and when it should be completed.

_____

_____

_____

_____

7. What type of fluids should be used in this drive train? Be specific.

_____

8. Locate the drain plug for the transmission/transaxle. Describe what it looks like and where you found it.

_____

_____

_____

9. Position the drain pan so it can catch all fluid from the drain plug.   Task Completed ☐

10. Allow all of the fluid to drain from the unit.   Task Completed ☐

11. Inspect the fluid in the drain pan for contamination and metal and other particles. Describe what you found.

    _____

    _____

12. What would cause the fluid to be contaminated with gold-color metal particles?

    _____

    _____

13. Reinstall the drain plug.   Task Completed ☐

14. Locate the filler plug and describe where it is located.

    _____

    _____

15. Remove the filler plug and fill the housing with the fluid until it reaches the level described by the manufacturer. How much fluid does the transmission/transaxle normally hold and how do you know when there is enough fluid in the housing?

    _____

    _____

    _____

    _____

    _____

16. Reinstall the filler plug after the correct fluid level is reached.   Task Completed ☐

Final Drive Service

17. Does the final drive unit require its own fluid, or is the fluid shared with the transmission?

    _____

    _____

18. If the final drive has its own fluid reservoir, locate the drain plug and describe what it looks like and where you found it.

    _____

    _____

    _____

19. Position the drain pan so it can catch all fluid from the drain plug.   Task Completed ☐

20. Allow all of the fluid to drain from the unit.   Task Completed ☐

21. Inspect the fluid in the drain pan for contamination and metal and other particles. Describe what you found.

_____

_____

22. Install the drain plug and then locate the filler plug and describe where it is located.

_____

_____

23. Remove the filler plug and fill the housing with the fluid until it reaches the level described by the manufacturer. How much fluid does the housing normally hold and how do you know when there is enough fluid in the housing?

_____

_____

_____

_____

_____

24. Reinstall the filler plug after the correct fluid level is reached.   Task Completed ☐

Problems Encountered

_____

_____

_____

Instructor's Comments

_____

_____

_____

MANUAL DRIVE TRAIN AND AXLES JOB SHEET 13

Troubleshoot a Clutch Assembly

Name _____ Station _____ Date _____

NATEF Correlation

This Job Sheet addresses the following **AST/MAST** task:

> **B.1.** Diagnose clutch noise, binding, slippage, pulsation, and chatter; determine necessary action.

Objective

Upon completion of this job sheet, you will be able to demonstrate the ability to troubleshoot a clutch assembly.

Tools and Materials

Droplight	Service information
Hoist	Socket set
Pry bar	Wheel chocks
Ruler	

Protective Clothing

Goggles or safety glasses with side shields

WARNING: *Be sure that the wheels are chocked properly and that the brake system is in good operating condition before beginning this task. Do not allow anyone to stand in front of or behind the vehicle during this test. Do not take any longer than necessary to determine if a problem exists.*

Describe the vehicle being worked on:

Year _____ Make _____ Model _____

VIN _____ Engine type and size _____

PROCEDURE

Check Clutch Chatter

1. Start the engine, set the parking brake, depress the clutch pedal fully, and shift the transmission into first gear. Increase the engine speed to about 1500 rpm and partially release the clutch pedal, just until the pressure plate first makes contact with the clutch disc, noting the clutch operation. Depress the clutch pedal and reduce the engine speed. Record the results on the Report Sheet for Clutch Troubleshooting, found at the end of this job sheet.

Task Completed ☐

2. Shift the transmission into reverse and repeat step 1. Record the results on the Report Sheet for Clutch Troubleshooting.

Task Completed ☐

3. If clutch chatter does not occur, increase the engine speed to about 2000 rpm and repeat steps 1 and 2. Record the results on the Report Sheet for Clutch Troubleshooting.

Task Completed ☐

4. If chatter occurs during the tests, raise the vehicle on a hoist. Check for loose or broken engine mounts, loose or missing bell housing bolts, and damaged linkage. Record the results on the Report Sheet for Clutch Troubleshooting. Correct any problems found during the inspection.

Task Completed ☐

5. Lower the vehicle and repeat steps 1–3.

Task Completed ☐

PROCEDURE

Check Clutch Slippage

1. Block the front wheels with wheel chocks and set the parking brake. Start the engine and run it for 15 minutes or until it reaches normal operating temperature.

Task Completed ☐

2. Shift the transmission into high gear and increase the engine speed to about 2000 rpm. Release the clutch pedal slowly until the clutch is fully engaged.

Task Completed ☐

   CAUTION: *Do not keep the clutch engaged for more than five seconds at a time. The clutch parts could become overheated and be damaged.*

3. If the engine does not stall, raise the vehicle on a hoist and check the clutch linkage. Correct any problems found during the inspection. Record the results on the Report Sheet for Clutch Troubleshooting.

Task Completed ☐

4. If any linkage problems were found and corrected, repeat steps 1 and 2. Record any problems in clutch operation on the Report Sheet for Clutch Troubleshooting.

Task Completed ☐

PROCEDURE

Check Clutch Drag

   NOTE: *The clutch disc and input shaft require about three to five seconds to come to a complete stop after engagement. This is known as "clutch spindown time." This is normal and should not be mistaken for clutch drag.*

1. Start the engine, depress the clutch pedal fully, and shift the transmission into first gear.

Task Completed ☐

2. Shift the transmission into neutral, but do not release the clutch pedal.

Task Completed ☐

3. Wait 10 seconds. Shift the transmission into reverse.

Task Completed ☐

4. If the shift into reverse causes gear clash, raise the vehicle on a hoist. Record the results on the Report Sheet for Clutch Troubleshooting.

Task Completed ☐

5. Check the clutch linkage and hydraulics.   Task Completed ☐

6. If any linkage or hydraulic problems were found and corrected, repeat   Task Completed ☐
   steps 1–3.

PROCEDURE

Check Pedal Pulsation

1. Start the engine. Slowly depress the clutch pedal until the clutch just   Task Completed ☐
   begins to disengage.

   NOTE: *A minor pulsation is normal.*

   Depress the clutch pedal further, and check for pulsation as the clutch
   pedal is depressed to a full stop. Record the results on the Report Sheet
   for Clutch Troubleshooting.

2. If no rubbing problems are found, remove one drive belt. Start the en-   Task Completed ☐
   gine and check for vibration. Shut off the engine. Repeat this process,
   removing the drive belts and checking for vibration after each drive belt
   is removed. If the vibration stops after a particular belt is removed, the
   problem is in the unit driven by that belt.

   CAUTION: *Do not run the engine for more than one minute when checking for a vibration with
   the belts removed.*

3. If a vibration is still present, check for a damaged crankshaft vibration   Task Completed ☐
   damper. Record the results on the Report Sheet for Clutch Trouble-
   shooting.

Problems Encountered

_____

_____

_____

Instructor's Comments

_____

_____

_____

Name _____ Station _____ Date _____

REPORT SHEET FOR CLUTCH TROUBLESHOOTING		
Pedal freeplay	Clutch pedal height specification	*Difference*
	Actual pedal height	
	Clutch pedal freeplay specification	*Difference*
	Actual pedal freeplay	
Clutch chatter test		*Yes*	*No*
	First gear 1500 rpm	
	Reverse gear 1500 rpm	
	First gear 2000 rpm	
	Reverse gear 2000 rpm	
Visual inspection		*Serviceable*	*Nonserviceable*
	Engine mounts	
	Transmission mounts	
	Transmission crossover	
	Bell housing bolt torque	
	Transmission to bell housing bolt torque	
	Clutch linkage and hydraulics	
	Linkage and hydraulic components	
	Oil leakage	
Clutch slippage test		*Yes*	*No*
	Engine stall	
Clutch drag test		*Yes*	*No*
	Gear clash	
Pedal pulsation test		*Yes*	*No*
	Top of pedal travel	
	Freeplay removed	
	Middle of travel	
	End of travel	
Conclusions and Recommendations _____		

_____

_____

MANUAL DRIVE TRAIN AND AXLES JOB SHEET 14

Clutch Linkage Inspection and Service

Name _____ Station _____ Date _____

NATEF Correlation

This Job Sheet addresses the following **AST/MAST** task:

> **B.2.** Inspect clutch pedal linkage, cables, automatic adjuster mechanisms, brackets, bushings, pivots, and springs; perform necessary action.

Objective

Upon completion of this job sheet, you will be able to inspect the clutch pedal linkage, cable, automatic adjuster mechanisms, brackets, bushings, pivots, and spring.

Tools and Materials

Lift

Service information

Protective Clothing

Goggles or safety glasses with side shields

Describe the vehicle being worked on:

Year _____ Make _____ Model _____

VIN _____ Engine type and size _____

Transmission type and number of forward speeds _____

PROCEDURE

1. Road-test the vehicle and describe the behavior of the clutch during all observed operations.

   _____

   _____

   _____

   _____

2. Inspect the following, indicating whether they appear to be okay or not.

	Okay	Should Be Replaced	Not There
Clutch Pedal Linkage	_____	_____	_____
Clutch Cable	_____	_____	_____
Automatic Adjuster Mechanisms	_____	_____	_____
Brackets	_____	_____	_____
Linkage Bushings	_____	_____	_____
Linkage Pivots	_____	_____	_____
Linkage Springs	_____	_____	_____
Clutch Master Cylinder	_____	_____	_____
Clutch Slave Cylinder	_____	_____	_____
Release Lever and Pivot	_____	_____	_____
Powertrain Mounts	_____	_____	_____

3. List the problems found in your diagnosis of this clutch.

_____

_____

_____

_____

4. List the specifications given in the service information for the following:

a. Pedal freeplay

_____

b. Pedal travel

_____

c. Clearance between cable and lever

_____

d. Clearance between slave cylinder push rod adjustment nut and wedge

_____

e. Torque for adjusting locknut

_____

5. Measure the same dimensions on the vehicle.

   a. Pedal freeplay

       _____

   b. Pedal travel

       _____

   c. Clearance between cable and lever

       _____

   d. Clearance between slave cylinder push rod adjustment nut and wedge

       _____

   e. Torque for adjusting locknut

       _____

6. Summarize the results of your measurements.

      _____

      _____

7. Check the tightness of the following bolts:

   a. Bell housing to engine block

       _____

   b. Transmission to bell housing

       _____

   c. Transmission to its mounts

       _____

8. Did you need to adjust or tighten any of these to specifications?

      _____

      _____

9. Summarize the preceding checks.

      _____

      _____

      _____

      _____

Problems Encountered

_____

_____

_____

Instructor's Comments

_____

_____

_____

MANUAL DRIVE TRAIN AND AXLES JOB SHEET 15

Clutch Inspection and Service

Name _____ Station _____ Date _____

NATEF Correlation

This Job Sheet addresses the following **AST/MAST** tasks:

B.3. Inspect and replace clutch pressure plate assembly, clutch disc, release (throw-out) bearing and linkage, and pilot bearing/bushing (as applicable).

B.6. Inspect flywheel and ring gear for wear and cracks; determine necessary action.

B.7. Measure flywheel runout and crankshaft end play; determine necessary action.

Objective

Upon completion of this job sheet, you will be able to inspect, measure, and replace the clutch disc, pressure plate, release bearing, pilot bushing or bearing, flywheel, and bell housing.

Tools and Materials

Bell housing alignment tool post assembly	Service information
Clutch alignment tool	Small mirror
Dial indicator	Straightedge
Surface plate	Feeler gauges
Micrometer	Pry bar
Vernier calipers or depth gauge	Pilot bearing puller
Flywheel turning tool	Transmission jack
HEPA vacuum made for removal of asbestos fibers	

Protective Clothing

Goggles or safety glasses with side shields

Describe the vehicle being worked on:

Year _____ Make _____ Model _____

VIN _____ Engine type and size _____

NOTE: *Before the clutch can be removed, both the transmissions/transaxle and the driveline must be removed from the vehicle. Refer to the appropriate job sheets for driveline and transmission/transaxle removal.*

PROCEDURE

Remove Clutch Assembly

1. Place fender covers over the fenders. Remove the battery negative cable.   Task Completed ☐

    WARNING: *When disconnecting the battery, disconnect the grounded terminal (usually the negative terminal) first. Disconnecting the positive terminal first can create a spark, which can cause the battery to explode and cause serious injury.*

2. Raise the vehicle on a hoist. Remove the starter and driveline.   Task Completed ☐

3. Remove the transmission following the appropriate procedures.   Task Completed ☐

4. Remove the hydraulic line and or slave cylinder as applicable clutch fork and cross-shaft linkage (if equipped) assemblies.   Task Completed ☐

5. Remove the bell housing. There are three basic bell-housing configurations commonly used.   Task Completed ☐

   a. The bell housing and transmission are removed as a unit. The bell housing may be part of the transmission or bolted to it.   Task Completed ☐

   b. The transmission is removed first. The bell housing is unbolted and removed from the back of the engine to expose the clutch assembly.   Task Completed ☐

   c. The transmission is removed, but the bell housing remains bolted to the engine. An access opening in the bell housing allows the technician to work on the clutch assembly.   Task Completed ☐

6. Remove any dust from the clutch assembly using an OSHA- and EPA-approved vacuum cleaner.   Task Completed ☐

   WARNING: *The dust inside the bell housing and on the clutch assembly contains asbestos fibers that are a health hazard. Do not blow this dust off with compressed air. Any dust should only be removed with a special vacuum cleaner designed for removal of asbestos fibers.*

7. To ensure proper reassembly, mark the flywheel and pressure plate clutch cover using a hammer and punch. Insert the clutch alignment tool through the clutch disc. Loosen and remove the bolts holding the clutch assembly to the flywheel.   Task Completed ☐

   CAUTION: *The pressure plate springs will push the pressure plate away from the flywheel. The bolts must be loosened a little at a time. If the bolts are loosened completely, one at a time, the pressure plate springs will push against the flywheel. This causes the clutch cover to become distorted or bent, ruining it.*

   WARNING: *The pressure plate assembly is heavy. Be sure the clutch alignment tool remains seated in the pilot bearing. Loosening the mounting bolts may unseat the alignment tool. The alignment tool may have to be tapped in frequently with a soft-faced mallet.*

8. Remove the pressure plate assembly from the vehicle.   Task Completed ☐

9. Remove the clutch alignment tool and clutch disc from the flywheel.   Task Completed ☐

PROCEDURE

For General Inspection

1. Perform a visual inspection of the components listed on the report sheet. Check for damage, wear, and warpage. Record your results on the report sheet found at the end of this job sheet.   Task Completed ☐

2. Locate the specification for flywheel lateral and radial runout on the Report Sheet for Clutch Inspection and Servicing.   Task Completed ☐

3. Mount the dial indicator on the bell housing or engine block. Place the tip of the indicator's plunger against the clutch surface of the flywheel. Check the flywheel for excessive lateral and radial runout. Record your results on the Report Sheet for Clutch Inspection and Servicing.

Task Completed ☐

4. Locate the specification for flywheel warpage. Record the specification on the Report Sheet for Clutch Inspection and Servicing.

Task Completed ☐

5. Using a straightedge and feeler gauge set, check the amount of flywheel warpage and taper. Record your results on the Report Sheet for Clutch Inspection and Servicing.

Task Completed ☐

6. If the flywheel has excessive runout, warpage, taper, light scoring, or light checking, remove the flywheel and resurface it.

Task Completed ☐

> **WARNING:** *Do not remove more than 10 percent of the flywheel's original thickness. A flywheel that is too thin can explode, causing serious injury or death.*

PROCEDURE

Check Face Runout

1. Locate the specification for face runout in the service information. Record the specification on the Report Sheet for Clutch Inspection and Servicing.

Task Completed ☐

2. Attach the clutch assembly to the flywheel.

Task Completed ☐

> **WARNING:** *The clutch assembly is heavy. If it falls, it can damage the assembly, or cause physical injury. Have a helper assist in holding and positioning the clutch assembly during installation.*

3. Install the bell-housing alignment tool post assembly through the bell-housing bore and into the clutch disc. Tighten the nut on the end of the post assembly until the clamp inside the clutch grips the clutch hub tightly.

Task Completed ☐

4. Install the dial indicator on the tool post assembly, and position it so the indicator's plunger contacts the face of the housing in a circular pattern just outside the housing bore.

Task Completed ☐

5. Gently pry the crankshaft rearward to eliminate all end play of the crankshaft. Hold the crankshaft in a rearward position. Set the dial indicator to zero.

Task Completed ☐

6. Rotate the crankshaft through one complete revolution, and check to see that the dial indicator returns to the zero position after one revolution.

Task Completed ☐

7. If the dial indicator did not return to zero, repeat steps 2–4.

Task Completed ☐

8. If the dial indicator returned to zero, rotate the crankshaft through two revolutions, making note of the highest reading obtained during the revolutions. Record your results on the Report Sheet for Clutch Inspection and Servicing.

Task Completed ☐

PROCEDURE

Check Bore Runout

1. Position the bell-housing alignment tool post assembly and dial indicator to check the bore's runout.
           Task Completed ☐

   NOTE: *You can use a rubber band to hold the dial indicator in place, but make sure it is just snug enough to provide a light pressure on the indicator's tip in the bore of the housing. If the rubber band is too tight, it may bind the dial indicator or distort the readings.*

2. Zero the dial indicator and rotate the crankshaft through one revolution, as in steps 5 and 6 of the procedure for checking face runout.
           Task Completed ☐

3. Make a note of the highest indicator reading through two complete crankshaft revolutions. Record your results on the Report Sheet for Clutch Inspection and Servicing.
           Task Completed ☐

Name _____ Station _____ Date _____

REPORT SHEET FOR CLUTCH INSPECTION AND SERVICING		
	Serviceable	*Nonserviceable*
Visual Inspection		
Oil leaks		
Flywheel		
Pilot bearing		
Clutch disc		
Pressure plate		
Finger height		
Release bearing		
Clutch fork		
Pressure plate cap		
Screws		

Item	Specifications	Actual	Service Required?
Flywheel lateral runout			
Flywheel radial runout			
Flywheel warpage			
Face runout			
Bore runout			
Conclusions and Recommendations _____			

_____

_____

PROCEDURE

Assemble Clutch

1. Using the service information, locate the bolt torque specifications required for the report sheet. Record the specifications on the Report Sheet for Clutch Reassembly and Installation found at the end of this job sheet.

   Task Completed ☐

2. Lubricate the pilot bearing or bushing according to the manufacturer's recommendation. Install the pilot bearing or bushing into the crankshaft flange.

   Task Completed ☐

3. Install any flywheel dowels.

   Task Completed ☐

4. Position the flywheel over the crankshaft mounting flange and screw in, by hand, two mounting bolts to hold the flywheel in position.

   Task Completed ☐

5. Insert the remaining crankshaft mounting bolts and finger-tighten them. Torque the flywheel mounting bolts in a crisscross pattern to the manufacturer's specifications.

   Task Completed ☐

   CAUTION: *Do not damage the surface of the flywheel around the dowel holes during installation.*

6. Place the clutch disc in its position against the flywheel. Line up the match marks on the flywheel and clutch cover, and finger-tighten two bolts through the pressure plate assembly to hold it in position.

   Task Completed ☐

   WARNING: *The clutch assembly is heavy. If it falls, it can damage the assembly, or cause physical injury. Have a helper assist in supporting and positioning the clutch assembly during reassembly.*

7. Select the correct size clutch alignment tool and insert it through the center of the pressure plate and into the clutch disc. Make sure the splines of the disc line up with the splines on the tool if the tool has them. Then, carefully push the tool through the disc and into the pilot bearing or bushing.

   Task Completed ☐

   CAUTION: *The clutch assembly-to-flywheel mounting bolts are specially hardened bolts. If it is necessary to replace any of these bolts because of damage, use only the recommended hardened bolts. Do not use ordinary bolts.*

8. Thread the rest of the pressure-plate mounting bolts and finger-tighten them. Torque the bolts in a crisscross pattern. Then, remove the alignment tool.

   Task Completed ☐

9. Clean the front and rear mounting surfaces of the bell-housing and mounting faces on the rear of the engine and the front of the transmission.

   Task Completed ☐

10. Install the bell-housing mounting bolts finger-tight. Torque them to the manufacturer's specifications.

    Task Completed ☐

11. Press the old throwout bearing out of the clutch fork. Press a new throwout bearing into the clutch fork.

    Task Completed ☐

12. Slip or clip the new throwout bearing to the clutch fork.

    Task Completed ☐

13. Put a small amount of the recommended lubricant inside the hub of the release bearing.

    Task Completed ☐

14. Install the dust cover or seal.

    Task Completed ☐

15. At this point, the transmission/transaxle and driveline can be reinstalled in the vehicle. See the job sheets that pertain to these procedures. After completing all of the reassembly procedures, road-test the vehicle and check for proper clutch operation.

Task Completed ☐

Problems Encountered

_____

_____

_____

Instructor's Comments

_____

_____

_____

Name _____ Station _____ Date _____

REPORT SHEET FOR CLUTCH REASSEMBLY AND INSTALLATION		
1. Torque specifications		
Flywheel to crankshaft		
Pressure plate to flywheel		
Bell housing to block		
Transmission to bell housing		
Transmission to mounts		
	Acceptable	_Not Acceptable_
2. Final road-test results		
Pedal free travel		
Pedal effort		
Pedal pulsation		
Clutch chatter		
Clutch slippage		
Clutch drag		
Vibrations		
Conclusions and Recommendations _____		
_____		
_____		

MANUAL DRIVE TRAIN AND AXLES JOB SHEET 16

Hydraulic Clutch Service

Name _____ Station _____ Date _____

NATEF Correlation

This Job Sheet addresses the following **MLR** tasks:

B.1. Check and adjust clutch master cylinder fluid level.

B.2. Check for system leaks.

This Job Sheet addresses the following **AST/MAST** tasks:

B.4. Bleed hydraulic clutch system.

B.5. Check and adjust clutch master cylinder fluid level; check for leaks.

Objective

Upon completion of this job sheet, you will be able to inspect the hydraulic lines and hoses for the clutch slave and master cylinder and bleed and adjust the system.

Tools and Materials

Clean fluid for the system   Machinist rule
Service information

Protective Clothing

Goggles or safety glasses with side shields

Describe the vehicle being worked on:

Year _____ Make _____ Model _____

VIN _____ Engine type and size _____

Transmission type and number of forward speeds _____

PROCEDURE

1. With your foot, slowly work the clutch pedal up and down. Pay attention to its entire travel. Check for any binding of the pedal linkage and/or a defective return spring. Describe your findings.

   _____

   _____

2. While working the pedal, does it feel like the hydraulic system is working properly?

   _____

   _____

3. Check the level and the condition of the fluid in the reservoir of the master cylinder. Describe your findings.

   NOTE: *Since brake fluid is generally used in the clutch hydraulic system, take precautions not to get the fluid in your eyes, or on the vehicle's painted areas.*

   _____

   _____

4. If the fluid level is low, fill the reservoir to the proper level.   Task Completed ☐

5. If the fluid was contaminated, the entire hydraulic system must be flushed   Task Completed ☐
   and bled after a thorough inspection of all of the parts in the system (If
   the hydraulic system is contaminated with mineral oil all rubber seals and
   lines will have to be replaced).

6. Carefully check the master cylinder, slave cylinder, hydraulic lines and
   hoses for evidence of leaks. Describe your findings.

   _____

   _____

7. If there was any evidence of a leak, find and repair the leak by replacing   Task Completed ☐
   the leaking part.

8. Anytime the system has been opened to make a repair, the system should   Task Completed ☐
   be bled. Bleeding may also be necessary if the system was allowed to be
   very low on fluid. Before bleeding, double-check the system for evidence
   of leaks and make sure the reservoir has the proper fluid level.

9. Check all mounting points for the master and slave cylinders. Make sure
   they do not move or flex when the pedal is depressed. Describe your
   findings.

   _____

   _____

10. Loosen the bleed screw on the slave cylinder approximately one-half turn.   Task Completed ☐

11. Fully depress the clutch pedal, and then move the pedal through three   Task Completed ☐
    quick and short strokes.

12. Close the bleeder screw immediately after the last downward movement   Task Completed ☐
    of the pedal. Then, release the pedal rapidly.

13. Recheck and correct the fluid level in the reservoir.   Task Completed ☐

14. Repeat steps 10, 11, and 12 until no air is evident in the fluid leaving the   Task Completed ☐
    bleeder screw.

15. Recheck and correct the fluid level in the reservoir.   Task Completed ☐

16. Measure clutch pedal free travel according to the manufacturer's recom-
    mendations (if adjustable). What were the results of this check?

    _____

    _____

17. Adjust the slave cylinder rod to set the clutch release bearing in its proper location and to set pedal free travel.

Task Completed ☐

18. With your foot, work the clutch pedal and describe how it feels now.

_____

_____

_____

Problems Encountered

_____

_____

_____

Instructor's Comments

_____

_____

_____

MANUAL DRIVE TRAIN AND AXLES JOB SHEET 17

Inspect and Adjust Shift Linkage

Name _____ Station _____ Date _____

NATEF Correlation

This Job Sheet addresses the following **AST/MAST** task:

C.1. Inspect, adjust, and reinstall shift linkages, brackets, bushings, cables, pivots, and levers.

Objective

Upon completion of this job sheet, you will be able to demonstrate the ability to inspect and adjust manual transmission and transaxle shift linkages.

Tools and Materials

Droplight	Service information
Hoist	Shifter alignment tool
Rags	Solvent

Protective Clothing

Goggles or safety glasses with side shields

Describe the vehicle being worked on:

Year _____ Make _____ Model _____

VIN _____ Engine type and size _____

PROCEDURE

Preparation

1. Place the shift lever in the neutral position. Raise the vehicle on a hoist. Wipe off all linkage parts with a rag.　　Task Completed ☐

2. Inspect the linkage for damage or wear. Check the shift-linkage rods for looseness and for worn or missing bushings. Record your results on the Report Sheet for Adjusting Floor-Mounted Shift Linkages.　　Task Completed ☐

3. Insert the shifter alignment tool to hold the linkage in its neutral position (if required).　　Task Completed ☐

PROCEDURE

Linkage Rods with Slotted Ends

1. Loosen the adjustment nut with an open-end or box-end wrench until the rod is free to move. Hold the slotted end using a screwdriver inserted into the slot. Loosen the adjustment nut with an open-end or box wrench.　　Task Completed ☐

2. Grasp the shifting arms and move them back and forth by hand, checking for smooth and positive movement.

Task Completed ☐

3. Place the shifting arms in their neutral position. When reconnecting the shift-linkage rods, adjust the rods to match the distance between the levers and arms. After the rods have been adjusted for length, tighten them in place.

Task Completed ☐

4. Tighten the adjustment nut by hand until the nut begins to contact the arm. Hold the slotted end of the rod by using a screwdriver inserted into the slot. Tighten the adjustment nut with an open-end or box-end wrench.

Task Completed ☐

PROCEDURE

Linkage Rods with Threaded Ends

1. Remove the holding pins from the swivels. Pull the swivels from the shift levers.

Task Completed ☐

2. Grasp the shifting arms and move them back and forth by hand, checking for smooth and positive movement.

Task Completed ☐

3. Place the shifting arms in their neutral position. When reconnecting the shift-linkage rods, the rods must be adjusted to match the distance between the levers and the arms. After the rods have been adjusted for length, tighten them in place.

Task Completed ☐

4. Lubricate the bushings with white grease. Rotate the adjustment swivels until the pin ends of the swivels slip easily into the bushings in the shift levers. Install new holding pins.

Task Completed ☐

NOTE: *Always use new holding pins. Old holding pins will break if they are reused.*

PROCEDURE

Linkage Rods with Clamps

1. Loosen the bolts that hold the clamps to the transmission arms. The bolt should be loosened until the rods are free to move.

Task Completed ☐

2. Grasp the shifting arms and move them back and forth by hand, checking for smooth and positive movement.

Task Completed ☐

3. Place the shifting arms in their neutral position. When reconnecting the shift-linkage rods, adjust them to match the distance between the levers and the arms. After the rods have been adjusted for length, tighten them in place.

Task Completed ☐

4. Tighten the clamp bolts with an open-end or box-end wrench.

Task Completed ☐

PROCEDURE

Final Checks—All Types

1. Remove the alignment tool.   Task Completed ☐

2. Lubricate the shift linkage with the proper lubricant.   Task Completed ☐

3. Lower the vehicle.   Task Completed ☐

4. Test the shift lever through all gears. The operation of the shift lever   Task Completed ☐
   should be smooth. Note any shifting noise or hard shifting problems.
   Record your results on the Report Sheet for Adjusting Floor-Mounted
   Shift Linkages.

5. Adjust the back drive rod.   Task Completed ☐

Problems Encountered

_____

_____

_____

Instructor's Comments

_____

_____

_____

REPORT SHEET FOR ADJUSTING FLOOR-MOUNTED SHIFT LINKAGES

	Serviceable	Nonserviceable
1. Inspection		
Linkage rods		
Bushings		
Shifting arms		
2. Final checks	Yes	No
Shift into all gears		
Operates smoothly		
Jumps out of gear		
Noises		
Hard shifting		
Conclusions and Recommendations _____		

MANUAL DRIVE TRAIN AND AXLES JOB SHEET 18

Electronic Controlled Transmissions

Name _____ Station _____ Date _____

NATEF Correlation

This Job Sheet addresses the following **MLR** task:

C.1. Describe the operational characteristics of an electronically controlled manual transmission/transaxle.

This Job Sheet addresses the following **AST/MAST** task:

C.2. Describe the operational characteristics of an electronically controlled manual transmission/transaxle.

Objective

Upon completion of this job sheet, you will be able to describe the operation of electronically controlled transmissions and transaxles.

Tools and Materials

Service information on the vehicle

Describe the vehicle being worked on:

Year _____ Make _____ Model _____

VIN _____ Engine type and size _____

PROCEDURE

1. Explain the differences between a manually shifted automatic transmission and an automatic manual transmission.

   _____

   _____

   _____

   _____

2. How does the manufacturer describe the automatic manual transmission you will be looking at?

   _____

   _____

3. What is the transmission called?

   _____

   _____

4. Briefly describe how the transmission works.

_____

_____

_____

_____

5. How many different shift modes are available and what are they called?

_____

_____

6. What does the manufacturer claim to be the benefits of this transmission?

_____

_____

_____

Instructor's Comments

_____

_____

_____

MANUAL DRIVE TRAIN AND AXLES JOB SHEET 19

Diagnosing Noise Problems

Name _____ Station _____ Date _____

NATEF Correlation

This Job Sheet addresses the following **MAST** tasks:

C.3. Diagnose noise concerns through the application of transmission/transaxle powerflow principles.

C.5. Diagnose transaxle final drive assembly noise and vibration concerns; determine necessary action.

Objective

Upon completion of this job sheet, you will be able to identify the cause of abnormal noises coming from the transmission or transaxle.

Tools and Materials

A vehicle with a manual transmission or transaxle

Protective Clothing

Goggles or safety glasses with side shields

Describe the vehicle being worked on:

Year _____ Make _____ Model _____

VIN _____ Engine type and size _____

PROCEDURE

1. Refer to the service information and describe what internal parts are rotating when the transmission/transaxle is in:

Neutral

_____

_____

First gear

_____

_____

Second gear

_____

_____

Third gear

_____

_____

Fourth gear

_____

_____

Fifth gear

_____

_____

Sixth gear

_____

_____

Reverse gear

_____

_____

2. Based on the preceding information, what parts are rotating in all gears?

_____

_____

3. What parts are rotating in all forward gears?

_____

_____

4. What parts only rotate when the unit is in reverse gear?

_____

_____

5. Describe the difference between a gear noise and a bearing noise.

_____

_____

6. Did the customer have a noise concern with the transmission/transaxle? If so, describe when it happens and what it sounds like.

_____

_____

7. Run the engine in neutral and then shift through all of the gears. Pay attention to when the noise gets louder and softer. Describe what is happening in the transmission when the noise changes.

   _____

   _____

8. Listen for an increase in noise as each gear is selected. What would cause this?

   _____

   _____

9. Raise the idle slowly in increments to 2000 rpm, and then to 2500 rpm. Does the noise decrease with an increase in engine speed? What are the most likely causes?

   _____

   _____

10. If there was noise while in neutral, does it go away when the vehicle is driven?

    _____

    _____

11. What could cause a grinding or clashing of gears when the gear ranges are changed?

    _____

    _____

12. Low lubricant levels or the use of the wrong type of fluid can cause what sort of noise? When will the noise be most evident?

    _____

    _____

13. What could be the cause of a noise or vibration that only occurs when the transmission is in gear?

    _____

    _____

14. What systems or components, other than the transmission/transaxle, could cause a noise or vibration that appears to be from the transmission?

    _____

    _____

Special Checks for Transaxles

1. List the components of a transaxle assembly that could cause noise/vibration whenever the vehicle is moving.

   _____

   _____

   _____

   _____

2. Most often a problem with the final drive assembly or the drive axles is the cause of abnormal noise. Use the following as a guideline for diagnosing those noises and then summarize your findings.

A knock at low speeds:

- Worn drive axle CV-joints
- Worn side gear hub counterbore

Noise most pronounced on turns:

- Differential gear noise

Clunk on acceleration or deceleration:

- Loose engine mounts
- Worn differential pinion shaft or side gear hub counterbore in the case
- Worn or damaged drive axle inboard CV-joints or worn axle splines

Clicking noise in turns:

- Worn or damaged outboard CV-joint

Noise from the final drive at all times:

- Incorrect adjustment of ring gear and pinion
- Ring gear or pinion damaged
- Damaged bearings on pinion shaft
- Damaged bearings in differential housing

Ticking sound while vehicle is moving:

- Ring and pinion gears are damaged
- Loose ring gear bolts

Gear chuckle:

- Loose ring gear bolts
- Insufficient amount of lubricant
- Damaged or missing differential thrust washers
- Side gears are loose in the case
- Damaged ring and/or pinion gears

Clunking noise when loads are changed:

- Excessive play in the differential
- Missing differential thrust washers

Unit whines while the vehicle is in motion:

- Incorrect adjustments of the ring and pinion gear set
- Worn axle bearings
- Insufficient amount of lubricant in the housing

Knocking noise:

- Loose ring gear bolts
- Damaged ring and pinion gear set

Unit makes a rumbling sound:

- Worn axle bearings
- Loose differential components

Vibrations caused by the differential not working freely on turns:

- Damaged or galled bearing surfaces between bevel gears and differential housing
- Damaged or galled bearing surfaces between bevel pinions and differential housing
- Damaged or galled bearing surfaces between bevel pinions and their shafts

Your conclusions:

_____

_____

_____

_____

Problems Encountered

_____

_____

_____

Instructor's Comments

_____

_____

_____

MANUAL DRIVE TRAIN AND AXLES JOB SHEET 20

Road-Test a Vehicle for Transmission Problems

Name _____ Station _____ Date _____

NATEF Correlation

This Job Sheet addresses the following **MAST** task:

C.4. Diagnose hard shifting and jumping out of gear concerns; determine necessary action.

Objective

Upon completion of this job sheet, you will be able to demonstrate the ability to properly road-test a vehicle to identify transmission problems.

Tools and Materials

A vehicle with a manual transmission or transaxle

Protective Clothing

Goggles or safety glasses with side shields

Describe the vehicle being worked on:

Year _____ Make _____ Model _____

VIN _____ Engine type and size _____

PROCEDURE

1. While driving the vehicle in town, obey the speed laws. Before beginning the road test, check the feel of the clutch pedal. Does it feel normal? Describe the feel and action of the clutch pedal.

   _____

   _____

2. Shift the transmission through the gears and identify any problem you might feel.

   a. Did it easily shift into first?   yes ___ no ___

   b. Did the gear change feel smooth?   yes ___ no ___

   c. Did it easily shift into second?   yes ___ no ___

   d. Did the gear change feel smooth?   yes ___ no ___

   e. Did it easily shift into third?   yes ___ no ___

   f. Did the gear change feel smooth?   yes ___ no ___

   g. Did it easily shift into fourth?   yes ___ no ___

   h. Did the gear change feel smooth?   yes ___ no ___

i. Did it easily shift into fifth?   yes ___ no ___

j. Did the gear change feel smooth?   yes ___ no ___

k. Did it shift easily into sixth?   yes ___ no ___

l. Did the gear change feel smooth?   yes ___ no ___

m. Did it easily shift into reverse?   yes ___ no ___

n. Did the gear change feel smooth?   yes ___ no ___

3. Describe what you think could be the cause of any shifting problem.

_____

_____

4. Now drive the vehicle, in one gear at a time. Accelerate, and then back off the throttle.

a. Did the transmission stay in first gear?   yes ___ no ___

b. Did the transmission stay in second gear?   yes ___ no ___

c. Did the transmission stay in third gear?   yes ___ no ___

d. Did the transmission stay in fourth gear?   yes ___ no ___

e. Did the transmission stay in fifth gear?   yes ___ no ___

f. Did the transmission stay in reverse gear?   yes ___ no ___

5. Describe what you think could be the cause of any jumping-out-of-gear problem.

_____

_____

_____

6. Describe the general condition of the transmission.

_____

_____

_____

Problems Encountered

_____

_____

_____

Instructor's Comments

_____

_____

_____

MANUAL DRIVE TRAIN AND AXLES JOB SHEET 21

Disassemble and Reassemble a Typical Transaxle

Name _____ Station _____ Date _____

NATEF Correlation

This Job Sheet addresses the following **MAST** task:

   C.6. Disassemble, clean, and reassemble transmission/transaxle components.

Objective

Upon completion of this job sheet, you will be able to disassemble and reassemble a typical manual transaxle.

Tools and Materials

Punch/drift set

Special tools as required

Service information

Protective Clothing

Goggles or safety glasses with side shields

Describe the vehicle being worked on:

Year _____ Make _____ Model _____

VIN _____ Engine type and size _____

PROCEDURE

1. Place the transaxle assembly into a suitable work stand. Secure the appropriate service information for this transaxle and use it as a guide through the following steps.

   Task Completed ☐

2. Loosen and remove all transaxle case-to-clutch housing attaching bolts. What did you need to remove in order to do this?

   _____

   _____

3. Separate the housing from the case. If the housing is difficult to loosen, tap it with a soft mallet.

   Task Completed ☐

4. Remove all external gearshift lever components. To do this, what else did you need to remove? Use a punch to drive the roll pin from the shift-lever shaft.

   _____

   _____

5. Grasp the input and main shafts and lift them as an assembly from the case. Note the position of the shift forks, so you will know where to place them when reinstalling them. Describe all you needed to do to pull the shafts out.

_____

_____

6. Remove the differential assembly from the case. Describe all you needed to do to pull the differential case out.

_____

_____

7. Remove the shift forks. What else did you need to remove?

_____

_____

8. Begin to remove the bearings and speed gears from the main shaft. These are normally removed in a particular order. Describe that order.

_____

_____

_____

9. Inspect the bearings and speed gears. Describe the condition of each.

_____

_____

_____

_____

10. Remove the synchronizer assemblies.   Task Completed ☐

11. Separate the synchronizer's hub, sleeve, and keys, noting their relative   Task Completed ☐
    positions and scribing their location on the hub and the sleeve prior to
    separation.

12. Inspect the synchronizer assemblies and describe each part for each assembly.

_____

_____

_____

_____

13. Identify all parts that need to be replaced during assembly. List them here.

_____

_____

_____

_____

14. Begin reassembly by lightly oiling the parts of the synchronizer.   Task Completed ☐

15. Assemble the synchronizer assemblies, being careful to align the index   Task Completed ☐
marks made during disassembly.

16. Install the synchronizer assemblies onto the main shaft.   Task Completed ☐

17. Install the speed gears and bearings onto the main shaft. Did the bearings
need to be pressed on?

_____

18. Install and tighten the shift fork assemblies.   Task Completed ☐

19. Install the differential assembly into the transaxle case.   Task Completed ☐

20. Place the main shaft control-shaft assembly on the main shaft so that the   Task Completed ☐
shift forks engage in their respective slots in the synchronizer sleeves.
Then, install the main shaft and input shaft assemblies.

21. Install all external gearshift lever components. To do this, what else did
you need to do?

_____

_____

22. Properly position and install the shift lever. What else needed to be
installed now?

_____

_____

23. Apply a thin bead of anaerobic sealant on the case's mating surface for the
clutch housing. What brand of sealer did you use?

_____

24. Install the clutch housing to the case.   Task Completed ☐

25. After the housing and case are fit snugly together, tighten the attaching
bolts to the specified torque. What is the recommended torque for the
bolts?

_____

_____

_____

_____

Problems Encountered

_____

_____

_____

Instructor's Comments

_____

_____

_____

MANUAL DRIVE TRAIN AND AXLES JOB SHEET 22

Drive Axle Inspection and Diagnosis

Name _____ Station _____ Date _____

NATEF Correlation

This Job Sheet addresses the following **AST/MAST** task:

    D.1. Diagnose constant-velocity (CV) joint noise and vibration concerns; determine necessary action.

Objective

Upon completion of this job sheet, you will be able to inspect and diagnose the front axles and joints on a FWD vehicle.

Tools and Materials

A FWD vehicle

Protective Clothing

Goggles or safety glasses with side shields

Describe the vehicle being worked on:

Year _____ Make _____ Model _____

VIN _____ Engine type and size _____

Describe the vehicle's general condition:

_____

CAUTION: *While driving the vehicle, make sure you obey all laws and have the permission of the owner to drive the vehicle.*

PROCEDURE

1. Raise the vehicle on a hoist.   Task Completed ☐

2. Carefully look at the CV joint boots on both drive axles and describe their condition. What is your recommendation?

    _____

    _____

    _____

3. Check the tightness of the boot clamps. Record your findings.

    _____

    _____

    _____

4. Visually inspect the shafts and describe their condition.

   _____

5. If there are no obvious problems with the boots, clamps, and shafts, lower the vehicle and prepare it for a road test.

   Task Completed ☐

6. Do a quick safety check of the vehicle. This should include the tires and lights. Describe your findings.

   _____

   _____

   _____

7. Drive the vehicle straight on the road. Pay attention to any unusual noises or handling problems. Describe your findings.

   _____

   _____

   _____

8. While moving straight at a low speed, accelerate hard and then let off the throttle. Pay attention to any unusual noises or handling problems. Describe your findings.

   _____

   _____

   _____

9. Now turn the vehicle to the right. Pay attention to any unusual noises or handling problems. Describe your findings.

   _____

   _____

   _____

10. Now turn the vehicle to the left. Pay attention to any unusual noises or handling problems. Describe your findings.

    _____

    _____

11. Return to the shop and record your conclusions from the road test. Include any recommended service.

    _____

    _____

    _____

Problems Encountered

_____

_____

_____

Instructor's Comments

_____

_____

_____

MANUAL DRIVE TRAIN AND AXLES JOB SHEET 23

Diagnosing RWD Noise and Vibration Concerns

Name _____ Station _____ Date _____

NATEF Correlation

This Job Sheet addresses the following **AST/MAST** tasks:

D.2. Diagnose universal joint noise and vibration concerns; perform necessary action.

D.5. Check shaft balance and phasing; measure shaft runout; measure and adjust driveline angles.

Objective

Upon completion of this job sheet, you will be able to road-check a RWD vehicle and identify noises and vibration concerns in the drive line, as well as inspect the components of the driveline, check and correct U-joint angles and driveshaft runout, and balance a driveshaft.

Tools and Materials

Brass drift   Dial indicator

Torque wrench   Transmission jack

Chalk, crayon, or paint stick   Inclinometer

Hose clamps   Service information

Protective Clothing

Goggles or safety glasses with side shields

Describe the vehicle being worked on:

Year _____ Make _____ Model _____

VIN _____ Engine type and size _____

Describe the vehicle's general condition:

_____

PROCEDURE

Customer Interview

1. If the customer has a concern about noise or vibration coming from the driveline, do a complete interview with him or her to identify the conditions of when the problem first occurs and/or continues to occur. Ask the customer the following questions:

   • What is the problem? Is it a shake, vibration, or noise?

   • Can you feel and hear the symptom?

- If the problem is a noise, what kind of noise is it?

- If the noise is a vibration or shake, where do you feel it?

- When does the noise or vibration occur?

- When did you first notice it?

- Were you going straight? Turning left? Turning right? Braking the car?

- What work has been done on the vehicle lately?

What did you find out during the interview?

_____

_____

2. What do you suspect as the cause of the noise or vibration based on the preceding interview?

_____

_____

_____

_____

PROCEDURE

Road Test

1. Attempt to operate the vehicle under the same conditions that the customer said the concern occurred. Were you able to do this?

_____

_____

2. Road-test the vehicle. It should be driven for enough time to warm it up. Then, accelerate the vehicle fairly slowly. Describe any abnormal noises or vibrations and when they occurred.

_____

_____

3. With the vehicle on a flat road and in steady traffic, maintain a constant speed. Describe any abnormal noises or vibrations and when they occurred.

_____

_____

4. Accelerate the vehicle quickly. Describe any abnormal noises or vibrations and when they occurred.

_____

_____

5. Let off the throttle and allow the vehicle to slow down. Describe any abnormal noises or vibrations and when they occurred.

_____

_____

6. Pay attention to all noises and vibrations during the road test. Which of the following were observed?

- Noise from the differential at all times   yes ___ no ___
- Rear wheel noise   yes ___ no ___
- A ticking sound while the vehicle is moving   yes ___ no ___
- The axle assembly chatters   yes ___ no ___
- A clunking noise when loads are changed   yes ___ no ___
- A growling sound when the vehicle is in motion   yes ___ no ___
- A whining sound when the vehicle is in motion   yes ___ no ___
- A knocking noise   yes ___ no ___
- A rumbling sound   yes ___ no ___
- A whistling noise   yes ___ no ___
- Groaning while the vehicle moves   yes ___ no ___
- Vibration during turns   yes ___ no ___
- Vibration during acceleration   yes ___ no ___
- Vibration during deceleration   yes ___ no ___
- Constant vibration while traveling at 35–45 mph   yes ___ no ___
- Constant vibration while traveling at 50–65 mph   yes ___ no ___

Based on the results of the preceding check, what could be the cause of the noise and vibration?

_____

_____

_____

_____

PROCEDURE

Visual Inspection

1. Place the transmission in neutral and raise the vehicle on a drive-on hoist. Check the fluid level in the rear axle. What did you find?

_____

_____

2. Check for leaks at the slip joint, U-joints, final drive pinion seal, and pinion companion flange. Describe your findings.

   _____

   _____

3. Check the vent cap on the axle housing to make sure it is open and not clogged or covered. What did you find?

   _____

   _____

4. Take a careful look at the exhaust system. Make sure nothing is loose or striking the chassis or other parts of the vehicle. What did you find?

   _____

   _____

5. Visually inspect the tires. Do they have any damage that would lead to noise or vibrations? Does the vehicle have snow tires? What did you find?

   _____

   _____

6. Check the wheels for distortion. What did you find? How did you check for this?

   _____

   _____

7. Check the wheel studs in the axle flanges. Make sure they are not broken or bent. Also check the condition of the studs and lug nuts. What did you find?

   _____

   _____

8. Shake and twist the driveshaft to locate worn or loose parts. Pry with a screwdriver around the U-joints. Describe your findings.

   _____

   _____

9. Check for dirt, undercoating, dents, or missing balancing weights on the driveshaft. Inspect the center-bearing rubber bushing and support bracket, if equipped. Describe your findings.

   _____

   _____

CAUTION: *Before attempting to check a center bearing, be sure the driving wheels and driveshaft are free to rotate.*

10. Check the center bearing for signs of wear or damage, if equipped. Describe your findings.

_____

_____

WARNING: *Extreme care should be taken when working around a rotating driveshaft. Severe injury can result from touching a moving shaft.*

PROCEDURE

Check U-Joint Angles

1. Locate the specification for U-joint angles in the service information. Clean the surfaces where the inclinometer will be mounted. What angle should the drive shaft be placed at?

_____

_____

CAUTION: *Do not force the inclinometer when setting it into position, otherwise a false reading will be recorded.*

2. Check the front U-joint angle and record the reading here.

_____

_____

3. Check the rear U-joint angle and record the reading here.

_____

_____

4. If necessary, correct the U-joint angles. How would you do that?

_____

_____

CAUTION: *Do not use too many shims. Measure at the center of each shim. It should be no thicker than ¼ inch (6.35 mm). If the rear U-joint angle is not correct, other problems may exist in the suspension. These problems include broken springs or an improperly placed spring seat.*

PROCEDURE

Check Driveshaft Runout

1. Locate the specification for driveshaft runout in the service information. Clean the areas on the driveshaft where the dial indicator plunger will ride. What is the maximum allowable runout?

_____

_____

2. Mount the dial indicator. Take runout readings at each end and at the center of the driveshaft. What were the readings?

_____

_____

3. If necessary, disconnect the driveshaft, rotate it 180 degrees on the differential companion flange, and reinstall it. Recheck the runout readings. What were the readings?

_____

_____

4. If necessary, replace the driveshaft. Recheck the runout readings. What were the readings?

_____

_____

PROCEDURE

Balance Driveshaft

1. Raise the vehicle on a hoist and support the rear axle housings.   Task Completed ☐

2. Remove the rear wheels. Replace and tighten the lug nuts with the flat edge against the brake drums.   Task Completed ☐

   WARNING: *Do not drive the rear axle without first locking the drums to the axles with the lug nuts. Failure to do so may result in a drum flying off of the axle and possibly causing severe injury.*

3. Clean the driveshaft thoroughly and determine the heavy spot on the driveshaft. Do this by very carefully holding a crayon or piece of chalk very near the spinning driveshaft. The first touch should revela the high spot of the driveshaft.   Task Completed ☐

   WARNING: *Keep the chalk or crayon away from the balancing weights on the driveshaft. Touching the balancing weights could cause serious injury.*

4. Add two hose clamps so that the screw part is on the light side of the drive-shaft. Recheck for vibration and noise. Never use more than two clamps.

Task Completed ☐

5. If necessary, move the hose clamps around the driveshaft to balance it. (Note: moving the heads of the screw clamps apart effectivity lightens the weight applied.)

Task Completed ☐

6. Recheck for vibration and noise.

Task Completed ☐

7. If necessary, disconnect the driveshaft, rotate it 180 degrees on its differential companion flange, and reinstall it.

Task Completed ☐

8. If necessary, move the hose clamps to the front of the driveshaft. Recheck for vibration and noise.

Task Completed ☐

9. If necessary, replace the driveshaft. Recheck for vibration and noise.

Task Completed ☐

10. Road-test the vehicle and make a note of any vibration and noise.

Task Completed ☐

Problems Encountered

_____

_____

_____

Instructor's Comments

_____

_____

_____

MANUAL DRIVE TRAIN AND AXLES JOB SHEET 24

Servicing FWD Wheel Bearings

Name _____ Station _____ Date _____

NATEF Correlation

This Job Sheet addresses the following **MLR** task:

D.1. Inspect, remove, and replace front-wheel drive (FWD) bearings, hubs, and seals.

This Job Sheet addresses the following **AST/MAST** task:

D.3. Inspect, remove, and replace front-wheel drive (FWD) bearings, hubs, and seals.

Objective

Upon completion of this job sheet, you will be able to replace front-wheel drive front-wheel bearings.

Tools and Materials

Lift	Brass drift
Torque wrench	Soft-faced hammer
Hydraulic press	CV joint boot protector
Various drivers and pullers	

Protective Clothing

Goggles or safety glasses with side shields

Describe the vehicle being worked on:

Year _____ Make _____ Model _____

VIN _____ Engine type and size _____

PROCEDURE

> **NOTE:** *This is a typical procedure. Check with the service information for the exact procedure you should follow and mark the differences in procedure as you progress through the following steps.*

1. Loosen the hub nut and wheel lug nuts. Have someone apply the brakes with the vehicle's weight on its tires. Loosen the axle hub nut and wheel lug nuts with the appropriate tools. Then, remove the axle hub nut.

   Task Completed ☐

2. Jack up the vehicle and remove the tire and wheel assembly. At this point, it's a good idea to install a boot protector over the CV joint boot to protect it while you are working.

   Task Completed ☐

3. Unbolt the front brake caliper, move it out of the way, and suspend it with wire. Make sure the flexible brake hose doesn't support the caliper's weight.

   Task Completed ☐

4. Remove the brake rotor.   Task Completed ☐

   NOTE: *Some vehicles use a sealed bearing assembly that can be unbolted and removed from the knuckle once the axle retaining nut has been removed.*

5. Loosen and remove the pinch bolt that holds the lower control arm to the   Task Completed ☐
   steering knuckle and separate the lower ball joint from the knuckle.

6. On some cars, an eccentric washer is located behind one or both of the   Task Completed ☐
   bolts that connect the spindle to the strut. These eccentric washers are
   used to adjust camber. Mark the exact location and position of this washer
   on the strut before you remove it. This allows you to reinstall the washer
   and the assembly without drastically disturbing the car's camber angle.

7. Once the washer is marked, remove the strut to knuckle retaining bolts.   Task Completed ☐

8. Remove the cotter pin and castellated nut from the tie rod end and pull   Task Completed ☐
   the tie rod end from the knuckle.

9. The tie rod end can usually be freed from its bore in the knuckle by using   Task Completed ☐
   a small gear puller or by tapping on the metal surrounding the bore in the
   knuckle.

10. Install the appropriate puller and press the hub-and-bearing assembly   Task Completed ☐
    from the half-shaft.

11. Pull the half-shaft away from the steering knuckle. Do not allow the CV   Task Completed ☐
    joint to drop downward while you remove the hub assembly. Support the
    half-shaft and the CV joint from the back side of the knuckle.

12. With the steering knuckle off, remove the snap ring or collar that retains   Task Completed ☐
    the wheel bearing.

13. Mount the knuckle on a hydraulic press so that it rests on the base of the   Task Completed ☐
    press with the hub free. Using a driver slightly smaller than the inside
    diameter of the bearing, press the bearing out of the hub. On some cars, this
    will cause half of the inner race to break out. Since half of the inner race
    is still on the hub, it must be removed. To do this, use a small gear puller.

14. Inspect the bearing's bore in the knuckle for burrs, score marks, cracks,   Task Completed ☐
    and other damage.

15. Lightly lubricate the outer surface of the new bearing assembly and the   Task Completed ☐
    inner bore of the knuckle.

16. Press the new bearing into the knuckle. Make sure all inner snap rings are   Task Completed ☐
    in place.

17. Install the outer retaining ring and bolt the brake splash shield onto the   Task Completed ☐
    knuckle.

18. Using the proper tool, press the new steering knuckle seal into place.   Task Completed ☐
    Lubricate the lip seal and coat the inside of the knuckle with a thin coat
    of grease.

19. Slide the knuckle assembly over the drive axle.   Task Completed ☐

20. Mount the knuckle to the strut and tighten the bolts. Make sure the   Task Completed ☐
    camber marks to the strut are positioned properly.

21. Reinstall the lower control arm ball joint and tie rod end to the steering knuckle. Then, mount the rotor and brake caliper. Torque all bolts to specifications. What are the specifications?

_____

22. Install the tire/wheel assembly and lower the vehicle.   Task Completed ☐

23. Once the car is set on the ground, tighten the new axle hub nut and wheel lug nuts to their specified torque. Install the wheel lug nuts and the new hub nut. What are the specifications?

_____

24. Lower the vehicle. Tighten all bolts to specifications. Stake or use a cotter pin to keep the hub nut in place after it is tightened. What are the specifications?

_____

25. Road-test the car and recheck the torque on the hub nut. Summarize the results of the road test.

_____

_____

26. Check the vehicle's wheel alignment. Although indexing the alignment   Task Completed ☐
during disassembly will get the alignment angles close, they won't be at the desired settings. Camber can be off by as much as three-quarters of a degree due to differences between the size of the camber holes and the bolt.

Problems Encountered

_____

_____

_____

Instructor's Comments

_____

_____

_____

MANUAL DRIVE TRAIN AND AXLES JOB SHEET 25

Servicing Universal Joints and CV Joints

Name _____ Station _____ Date _____

NATEF Correlation

This Job Sheet addresses the following **MLR** task:

D.2. Inspect, service, and replace shafts, yokes, boots, and universal/CV joints.

This Job Sheet addresses the following **AST/MAST** task:

D.4. Inspect, service, and replace shafts, yokes, boots, and universal/CV joints.

Objective

Upon completion of this job sheet, you will be able to remove, inspect, and install both outer and inner CV joints.

Tools and Materials

Brass drift   Service information

Soft-jaw vise   Solvent tank or other cleaning equipment

Protective Clothing

Goggles or safety glasses with side shields

Describe the vehicle being worked on:

Year _____ Make _____ Model _____

VIN _____ Engine type and size _____

Describe the vehicle's general condition:

_____

PROCEDURE

Remove and Inspect Outer Joint

1. Remove the drive axle half-shaft from the vehicle, following the procedure in the service information.

   Task Completed ☐

2. Secure the half-shaft in a soft-jaw vise. Mark the location of the boot to the shaft. Cut off the boot clamps and boot from the outer joint. Wipe away all grease to expose the snap ring or circlip.

   Task Completed ☐

3. If the joint is retained by a circlip, remove the joint from the shaft by striking the outer housing with a soft-faced hammer.

   Task Completed ☐

4. If the joint is retained by a snap ring, hold the snap ring open with snap ring pliers and remove the joint from the shaft by striking the outer housing with a soft-faced hammer.

Task Completed ☐

   NOTE: *If the joint refuses to move, use a brass drift against the face of the inner joint race to drive the joint off the shaft.*

   WARNING: *Always wear safety goggles or glasses with side shields when using a hammer and punch.*

5. Secure the outer housing in the vise to make working easier. Do not damage the splines in the outer housing end shaft with vise pressure.

Task Completed ☐

6. Tilt the inner race and remove the joint balls in the sequence described.

Task Completed ☐

7. Tilt the inner race at a 90-degree angle to the outer housing. Align the cage windows with the outer race, and then lift and remove the cage and inner race cross from the housing. Rotate the inner race upward and out of the cage.

Task Completed ☐

8. Clean all components.

Task Completed ☐

   NOTE: *After cleaning the constant velocity joint with solvent to inspect it, be sure to let the solvent dry completely before lubricating the reassembled joint. If the solvent is still wet, it might contaminate the lubricant and cause the reassembled CV joint to fail prematurely. It is also suggested that the CV joint be rinsed with a CV joint cleaner to remove residue completely.*

9. Inspect the joint for scoring or excessive wear on the bearings, cage windows, and inner and outer bearing races. Replace the entire joint if excessive wear is found on the bearings or grooves. Record the results on the Report Sheet for Inspection of Outer Constant Velocity Joints found at the end of this job sheet.

Task Completed ☐

10. Examine the joint for cracks, chips, or brinelling.

Task Completed ☐

    NOTE: *Brinelled cages are the main cause of clicking. A new cage with proper dimensions will normally correct this problem, but replacement of the entire wheel joint assembly is recommended if either condition exists. Shiny areas on the races and cages normally do not indicate a need for replacement unless the areas are excessively worn.*

    Record the results on the Report Sheet for Inspection of Outer Constant Velocity Joints.

Task Completed ☐

PROCEDURE

Pack and Install Joint

1. If the components pass inspection, carefully reassemble the joint. Apply a light coat of oil to all parts before reassembly.

Task Completed ☐

2. Align the inner race with the cage window and rotate it downward into the cage. Position the cage windows between the ball races and rotate them downward.

Task Completed ☐

3. Swing the race cross and cage 90 degrees into the housing with the larger counterbore of the race cross facing outward.

Task Completed ☐

4. Evenly distribute the grease provided in the repair kit into all ball bearing grooves.

Task Completed ☐

5. Install as much grease as possible through the inner race and splined housing. Make sure that grease comes out between the outer race, cage, and bearings. Then, install the rest of the grease into the boot. Use all of the grease supplied in the kit.

Task Completed ☐

6. Tilt the cage and race cross (with the bearing grooves and windows aligned) toward the ball grooves in the housing, and insert the first ball. Insert the second ball the same way, but in the opposite side. Install the remainder of the balls in a similar fashion.

Task Completed ☐

7. Loosely place the new boot clamp on the half-shaft. Carefully place the boot over the spline onto the shaft. Put electric tape over the splines to prevent boot damage during installation.

Task Completed ☐

8. Install a new circlip. Position the joint on the shaft splines. Install the joint onto the shaft. Using a soft mallet, sharply tap the joint onto the splined half-shaft over the circlip. Pull lightly on the joint to make sure the circlip is seated correctly.

Task Completed ☐

9. Install the boot and small clamp on the half-shaft. Use the remainder of the special grease to pack the boot.

Task Completed ☐

10. With the large boot end properly placed on the joint housing, make sure no twists or crinkles appear on the boot. Install and tighten both boot clamps.

Task Completed ☐

   NOTE: *The boot clamps on some vehicles must be tightened to a specified torque. Check the service manual before tightening.*

   CAUTION: *Do Not Overtighten.*

11. Flex the joints through a full range of motion to be sure they work smoothly. The complete half-shaft is now ready for installation.

Task Completed ☐

PROCEDURE

Remove and Inspect the Inner Joint

1. With the axle removed from the vehicle: Turn the axle shaft so the inner joint can be worked on.

Task Completed ☐

2. Push the housing to compress the retention spring lightly inside the joint. Bend the retaining tabs back using pliers. Hold the housing as the spring pushes it from the tripod. Separate the tripod and housing carefully.

Task Completed ☐

3. Remove the snap ring. Do not allow the bearings to fall off the tripod. Use a wide elastic rubber band or tape to hold the bearings in place on the tripod. The tripod can be removed by hand or by tapping gently with a brass drift or wood block.

Task Completed ☐

   WARNING: *Always Wear Safety Goggles Or Glasses With Side Shields When Using A Hammer And Punch.*

4. Wipe any grease from the constant velocity (cv) joint bearings and bearing races.

Task Completed ☐

> **CAUTION:** *Do not clean with solvent. Solvent could remain inside the needle bearing pockets and cause failure during operation.*

Inspect the rest of the assembly for internal damage. Examine the fit between the rollers and housing. Excessive free play, roughness on either the roller or track surfaces, or damage to the bearings or trunnion call for joint replacement. Record the results on the Report Sheet for Inner Constant velocity (CV) Joint.

5. Inspect the grease inside the housing. A gritty texture signals dirt penetration and possible damage. Clean the inside of the housing with rags.

Task Completed ☐

PROCEDURE

Joint Installation

> **NOTE:** *The replacement joint illustrated in this procedure is a spring-loaded design that does not have a retaining circlip to hold it in the differential.*

1. Wipe any remaining grease from the half-shaft splines and clean the shaft boot seating surface with solvent. Let the shaft dry completely. Temporarily tape the splines to avoid damaging the boot.

Task Completed ☐

2. Slide the new boot onto the half-shaft and align the boot with the mark made before removing the old one. Install the boot clamp. Bend over the clamp and cut off all but one-half inch or so. Bend this small tap under itself with pliers, and lightly tap it secure. Remove the protective tape from the shaft.

Task Completed ☐

3. Tape the rollers securely on the new tripod so they will not accidentally slide off the tripod during assembly. Place the tripod on the shaft splines with the chamfered or scribed end facing toward the outboard joint end.

Task Completed ☐

4. Install the retaining ring into the shaft groove to lock the tripod assembly onto the shaft. Check that the ring is securely seated by pulling out the joint.

Task Completed ☐

5. Pack the roller faces of the tulip housing raceway with the high-temperature grease contained in the replacement kit. Do not substitute any other type of grease.

Task Completed ☐

6. Place the spring cup into the housing spring pocket, and lightly grease the cup. Reinstall the tulip housing. Slip the new retaining ring onto the shaft. Carefully remove the tape from the rollers and slide the tulip housing onto the tripod. Check that the spring remains in the pocket and that the cup connects with the round end of the connecting shaft.

Task Completed ☐

7. Compress the spring by hand so the housing seats in the retaining ring. Use the properly sized C-clamp to hold the joint in a compressed position until the retaining ring can be secured.

Task Completed ☐

8. To secure the retaining ring, remove the joint from the vise and use a hammer and punch to tightly compress the retaining ring to the housing. Work completely around the ring to ensure a tight fit.

Task Completed ☐

9. Place the shaft back into the vise and position the boot on the tulip housing, making certain it is not dimpled or otherwise deformed. Remove any irregularities on the boot by working it between your fingers or by venting air into the boot with a dull screwdriver.

Task Completed ☐

10. Install the large boot clamp and tighten it. Be careful not to cut or damage the boot or clamp as you work.

Task Completed ☐

PROCEDURE

Universal-Joint Replacement

1. Mark the position of the driveshaft on the differential flange so the driveshaft can be reinstalled in the same position.

Task Completed ☐

2. Remove the driveshaft, and plug the transmission output shaft to prevent loss of transmission fluid.

Task Completed ☐

3. Determine how the universal joints are retained in the driveshaft. Some are held in with snap rings on the top or bottom of the bearing caps, while others are retained by injected plastic. Describe how the end caps are retained on your vehicle and the recommended method of removal.

_____

_____

_____

4. Using a U-joint press, press the bearing caps through the driveshaft. (It is best to use a press instead of hammering on the driveshaft to remove the caps.) Describe how you removed the U-joints.

_____

_____

5. Describe the condition of the bearings and trunnions.

_____

_____

_____

_____

_____

6. Make sure to keep the new universal joints clean. Also, make sure that the needle bearings do not fall out of the bearing caps while handling them. If the original bearings were retained by injected plastic, make sure the snap ring grooves are free of plastic. Describe the process to reinstall the new universal joint.

_____

_____

_____

Problems Encountered

_____

_____

_____

Instructor's Comments

_____

_____

_____

REPORT SHEET FOR INSPECTION OF OUTER CONSTANT VELOCITY JOINTS		
	Serviceable	_Nonserviceable_
Visual Inspection		
Balls		
Cage		
Inner race		
Outer race		
Conclusions and Recommendations _____		

REPORT SHEET FOR INNER CV JOINT

	Serviceable	*Nonserviceable*
Inspection		
Rollers		
Needle bearings		
Trunnion		
Outer "tulip" cage		
Conclusions and Recommendations _____		

_____

_____

MANUAL DRIVE TRAIN AND AXLES JOB SHEET 26

Drive Axle Leak Diagnosis

Name _____ Station _____ Date _____

NATEF Correlation

This Job Sheet addresses the following **MLR** task:

E.1. Clean and inspect differential housing; check for leaks; inspect housing vent.

This Job Sheet addresses the following **AST/MAST** task:

E.1.1. Clean and inspect differential housing; check for leaks; inspect housing vent.

Objective

Upon completion of this job sheet, you will be able to identify the cause of drive axle fluid leakage problems, be able to drain and refill the differential housing, and know how to check and adjust the differential fluid levels.

Tools and Materials

Service information

Fluid for the differential

Limited slip additive as needed

Protective Clothing

Goggles or safety glasses with side shields

Describe the vehicle being worked on:

Year _____ Make _____ Model _____

VIN _____ Engine type and size _____

PROCEDURE

1. To identify the exact source for a leak, a careful inspection must be completed. Look at the drive axle and note the spots where there may be a leak.

   _____

   _____

2. Check the area around the pinion shaft. An improperly installed or damaged drive pinion seal will allow the lubricant to leak past the outer edge of the seal. Any damage to the seal's bore, such as dings, dents, and gouges, will distort the seal casing and allow leakage. Also, the spring that holds the seal lip against the companion flange may be knocked out and

allow leakage past the seal's lip. Clean the area around the leak so you can determine the location of the leak and record your findings.

_____

_____

3. It is also possible for oil to leak past the threads of the drive pinion nut or the pinion retaining bolts. Removing the nut or bolts, applying thread or gasket sealer on the threads, and torquing the nuts or bolts to specifications can stop these leaks. Clean the area around the leak so you can determine the location of the leak and record your findings.

_____

_____

4. Leakage past the stud nuts that hold the carrier assembly to the axle housing can be corrected by installing copper washers under the nuts. Always make sure there is a copper washer under the axle ID tag. Clean the area around the leak so you can determine the location of the leak and record your findings.

_____

_____

5. Check the housing cover for signs of leakage. Check for poor gasket installation, loose retaining bolts, or a damaged mating surface. Most late-model vehicles do not use a gasket on the housing cover; instead, silicone sealer is used. The old sealant should be cleaned off before applying a new coat to the surface. If a housing cover is leaking, inspect the surface for imperfections, such as cracks or nicks. File the surface true and install a new gasket. If the surface cannot be trued, apply some gasket sealer to the surface before installing the new gasket. Always follow the recommended tightening sequence and torque specifications when tightening an axle housing cover. Describe your findings.

_____

_____

6. Check for lubricant leaks on the housing itself. At times, lubricant will leak through the pores of the housing. To correct this problem when the porous area is small, force some metallic body filler into the area. After the filler has set, seal the area with an epoxy-type sealer. If the leaking pore is rather large, drill a hole through the area and tap an appropriately sized setscrew into the hole, then cover the area with an epoxy sealer. Describe your findings.

_____

_____

7. Check the level of the fluid in the assembly. Describe your findings and state what type of fluid is recommended for this vehicle.

_____

_____

_____

8. Check the vent for signs of leakage. If the wrong vent was installed in the axle housing or if there is an excessive amount of lubricant or oil turbulence in the axle, lubricant may leak through the axle vent hose. Also check for a crimped or broken vent hose. If the cause of the leakage is an overfilled axle assembly, drain the housing and refill the unit with the specified amount and type of lubricant. Describe your findings.

_____

_____

9. Check the brake backing plates and drums for signs of fluid. If fluid is present, it is usually caused by a worn or damaged axle seal. However, if the seal's bore in the axle tube is damaged, the seal will be unable to seal properly. Whenever fluid is leaking past the axle seal, the bore should be checked and the seal replaced. Describe your findings.

_____

_____

10. Some drive axle units are fitted with an ABS sensor. Lubricant can leak from around the sensor's O-ring if it is damaged. Check it and describe your findings.

_____

_____

Problems Encountered

_____

_____

_____

Instructor's Comments

_____

_____

_____

MANUAL DRIVE TRAIN AND AXLES JOB SHEET 27

Differential Housing Service

Name _____ Station _____ Date _____

NATEF Correlation

This Job Sheet addresses the following **MLR** tasks:

E.2. Check and adjust differential housing fluid level.

E.3. Drain and refill differential housing.

This Job Sheet addresses the following **AST/MAST** tasks:

E.1.2. Check and adjust differential housing fluid level.

E.1.3. Drain and refill differential housing.

Objective

Upon completion of this job sheet, you will be able to inspect and flush a differential housing and refill it with the correct lubricant.

Tools and Materials

Basic hand tools

Clean rag

Protective Clothing

Goggles or safety glasses with side shields

Describe the vehicle being worked on:

Year _____ Make _____ Model _____

VIN _____ Engine type and size _____

Transmission type and number of forward speeds _____

PROCEDURE

1. On a RWD differential and final drive assembly, check the fluid level with the vehicle on a level surface and the axle at normal operating temperature. The fluid level should be even with the bottom of the fill plug opening in the axle housing unless otherwise specified. State the type of recommended fluid.

   _____

   _____

2. If the fluid level is low, inspect the differential and axle housing for signs of leakage. What did you find?

_____

_____

3. State the condition of the fluid and explain what you think may have caused the unusual smell or the contamination.

_____

_____

4. On most FWD models with manual transaxles, the transmission and final drive assembly are lubricated with the same fluid and share the same fluid reservoir. State the type of recommended fluid.

_____

_____

5. Check the fluid level at the transaxle fill plug or dipstick. The fluid should be even with the bottom of the fill plug opening unless otherwise specified. What did you find?

_____

_____

6. State the condition of the fluid and explain what you think may have caused the unusual smell or the contamination.

_____

_____

7. If the fluid was contaminated and the problem was not caused by the destruction of parts in the differential, the housing should be flushed. To do this, drain the fluid from the housing.   Task Completed ☐

8. Refill the housing with clean fluid and rotate the wheels and/or drive shaft to circulate the fluid.   Task Completed ☐

9. Drain this fluid and refill the housing with a fresh supply of fluid.   Task Completed ☐

10. Rotate the wheels and/or drive shaft several times, and then reexamine the condition of the fluid. If the fluid still has traces of residue, flush the housing again or until the fluid is clean.   Task Completed ☐

11. Is the differential a limited-slip unit? If so, what fluid is recommended and what additive should you mix with the fluid? Before adding friction modifiers to the lubricant of a limited-slip differential, always check its compatibility. Using the wrong additive may cause the axle's lubricant to break down, which would destroy the axle very quickly.

_____

_____

Problems Encountered

_____

_____

_____

Instructor's Comments

_____

_____

_____

MANUAL DRIVE TRAIN AND AXLES JOB SHEET 28

Diagnosing Differential Noise and Vibration Concerns

Name _____ Station _____ Date _____

NATEF Correlation

This Job Sheet addresses the following **MAST** task:

E.1.4. Diagnose noise and vibration concerns; determine necessary action.

Objective

Upon completion of this job sheet, you will be able to road-check a RWD vehicle and identify noises and vibration concerns from differential and related parts.

Tools and Materials

Service information

Protective Clothing

Goggles or safety glasses with side shields

Describe the vehicle being worked on:

Year _____ Make _____ Model _____

VIN _____ Engine type and size _____

PROCEDURE

1. Is the concern a noise or vibration? A noise can be as simple as loose trim, or two parts in close proximity, or it can indicate a more serious problem such as a damaged bearing. A vibration usually means an imbalance problem and occurs at a very specific speed.

   _____

   _____

   _____

2. When does the noise or vibration occur? If the concern is a noise, does it change on different types of pavement (concrete verses asphalt)?

   _____

   _____

   _____

3. A differential may make noise based upon the load on the gear set, such as coasting, maintaining a certain speed (floating) acceleration, or deceleration. Is the vehicle experiencing any of these symptoms? If so, describe the noise.

_____

_____

_____

4. Check the condition and level of the fluid in the differential. Record what you found.

_____

_____

5. Are there any signs of damage to the tires or wheels that may cause the noise or vibration concern?

_____

_____

6. If the problem was a vibration, especially at speeds under 40 mph, check the universal joints for wear or looseness. Describe what you found.

_____

_____

7. If the problem is a vibration over speeds of about 40 mph, recheck the balance of the wheels. Describe what you found.

_____

_____

8. If the noise is a rumbling sound from the differential that might change during turns (depending on the severity of the damage), what do you think is a possible source of the noise, based on the design of a differential?

_____

_____

9. If the noise from the differential is a constant whirling sound that does not change during turns, what do you think could be a possible cause of the problem, based on the design of a differential?

_____

_____

10. If the differential makes a whining noise that changes or becomes more pronounced during acceleration, coasting, floating, or deceleration, a possible cause can be an incorrect gear setup, ring and pinion wear, or wear in the differential that may influence the relationship of the

ring gear and pinion. Has the differential undergone any repairs or replacement of parts? Describe the noise problem, when it occurs and a possible solution for it.

_____

_____

_____

11. Were you able to determine the source of the noise/vibration from the vehicle? If so, list the cause here. If not, list the next step you plan to take.

_____

_____

_____

Problems Encountered

_____

_____

_____

Instructor's Comments

_____

_____

_____

MANUAL DRIVE TRAIN AND AXLES JOB SHEET 29

Companion Flange and Pinion Seal Service

Name _____ Station _____ Date _____

NATEF Correlation

This Job Sheet addresses the following **AST** task:

E.1.4. Inspect and replace companion flange and pinion seal; measure companion flange runout.

This Job Sheet addresses the following **MAST** task:

E.1.5. Inspect and replace companion flange and pinion seal; measure companion flange runout.

Objective

Upon completion of this job sheet, you will be able to inspect and replace the companion flange and pinion seal. You will also be able to measure companion flange runout.

Tools and Materials

Pinion flange holding tool	Seal remover
Catch pan	Seal puller
Center punch	Inch-pound torque wrench
Slide hammer	Service information
Fresh lubricant	

Protective Clothing

Goggles or safety glasses with side shields

Describe the vehicle being worked on:

Year _____ Make _____ Model _____

VIN _____ Engine type and size _____

PROCEDURE

1. If the pinion seal is a source of fluid leakage, it should be replaced. To replace a pinion seal, remove the drive shaft from the companion flange.　　Task Completed ☐

2. Remove the flange nut. You may need a special wrench to hold the flange while removing the nut.　　Task Completed ☐

3. Pull the companion flange off the pinion splines.　　Task Completed ☐

4. Inspect the sealing surface of the flange to make sure it is smooth and free of burrs. What did you find?

_____

_____

5. With a slide hammer, remove the pinion seal.   Task Completed ☐

6. Before you install a new seal, lubricate the seal's lip and coat the outside   Task Completed ☐
diameter of the seal with gasket sealer. Use a seal driver to install the new
seal flush against the carrier or axle housing. Make sure the spring around
the seal lip remains in place.

7. After the seal is installed, lubricate the seal area of the companion flange,   Task Completed ☐
and push it carefully into the seal, making sure the seal isn't damaged.

8. Start a new pinion flange nut onto the threads and set bearing preload.
The tightness of the pinion flange nut is critical because it determines the
pinion bearing preload. If the flange nut is overtightened, the pinion bearing preload must be readjusted. Describe the recommended procedure for
setting the bearing preload.

_____

_____

_____

_____

9. Excessive companion flange runout can cause a number of problems,   Task Completed ☐
including wear on the seal and driveline vibrations. To check flange
runout, mount a dial indicator on the axle housing so that the indicator's
plunger rests on the outside face of the flange.

10. Mark the spot where the indicator's plunger contacts the flange.   Task Completed ☐

11. Set the indicator to zero and rotate the wheels to turn the flange. Rotate   Task Completed ☐
the flange one complete turn.

12. The total amount of needle deflection on the indicator is the total amount
of runout. What was the measured runout?

_____

_____

13. Compare your reading to the specifications and summarize the results.

_____

_____

_____

_____

14. Once the flange nut is installed and properly tightened, install the drive   Task Completed ☐
shaft.

15. Check the fluid in the differential, and then road-test the vehicle and again   Task Completed ☐
check for leaks.

Problems Encountered

_____

_____

_____

Instructor's Comments

_____

_____

_____

MANUAL DRIVE TRAIN AND AXLES JOB SHEET 30

Measure and Adjust Pinion Depth, Bearing Preload, and Backlash

Name _____ Station _____ Date _____

NATEF Correlation

This Job Sheet addresses the following **MAST** tasks:

E.1.6. Inspect ring gear and measure runout; determine necessary action.

E.1.7. Remove, inspect, and reinstall drive pinion and ring gear, spacers, sleeves, and bearings.

E.1.8. Measure and adjust drive pinion depth.

E.1.9. Measure and adjust drive pinion bearing preload.

E.1.10. Measure and adjust side bearing preload and ring and pinion gear total backlash and backlash variation on a differential carrier assembly (threaded cup or shim types).

E.1.11. Check ring and pinion tooth contact patterns; perform necessary action.

E.1.12. Disassemble, inspect, measure, and adjust or replace differential pinion gears (spiders), shaft, side gears, side bearings, thrust washers, and case.

Objective

Upon completion of this job sheet, you will be able to check and adjust drive pinion depth, bearing preload, and ring and pinion gear backlash.

Tools and Materials

Service information   Dial indicator

Pinion gear flange holding tool   Pinion depth gauge

Micrometer   Torque wrench

Protective Clothing

Goggles or safety glasses with side shields

Describe the vehicle being worked on:

Year _____ Make _____ Model _____

VIN _____ Engine type and size _____

PROCEDURE

Drive Pinion Depth

1. Check the condition of the pinion bearings; replace them if necessary.   Task Completed ☐

2. Check the pinion gear for any markings indicating additional adjustments. Record the markings:

_____

3. Set up the pinion depth gauge according to the procedure outlined in the service information.   Task Completed ☐

4. Set up the dial indicator on the carrier housing.   Task Completed ☐

5. Make the necessary readings with the indicator and record the results:

   _____

6. How much needs to be added or subtracted to achieve proper pinion depth? _____

7. Refer to the service information to determine the correct size of pinion shim that should be used.   Task Completed ☐

8. Install the shim and bearing on the pinion gear shaft.   Task Completed ☐

9. Install the pinion gear into the carrier housing.   Task Completed ☐

PROCEDURE

Pinion Bearing Preload

1. Install the pinion gear, crush sleeve, and bearing into the carrier housing.   Task Completed ☐

2. Install the pinion seal into the housing.   Task Completed ☐

3. Install the pinion flange, washer, and nut on the pinion.   Task Completed ☐

4. Using the flange holding tool, tighten the pinion nut.   Task Completed ☐

5. Using a torque wrench, measure the torque required to turn the pinion gear. Required torque is: _____

6. Refer to the service information for the proper torque required to turn the pinion. Specified torque is: _____

7. Tighten the pinion nut until the proper torque reading is reached.   Task Completed ☐

PROCEDURE

Ring and Pinion Gear Backlash

1. Check the ring gear for runout by setting the dial indicator on the back side of the ring gear.   Task Completed ☐

2. Rotate the ring gear one complete revolution and note the movement on the dial indicator. Describe what you observed.

   _____

   _____

3. What was the highest reading? _____

4. What was the lowest reading? _____

5. Subtract the lowest from the highest. This indicates the total runout of the ring gear. What was it? _____

6. If the runout was not within specifications, check the runout of the carrier before replacing the ring gear.   Task Completed ☐

7. Now, install the differential case and ring gear into the carrier housing.   Task Completed ☐

8. Mount the dial indicator onto the carrier housing.   Task Completed ☐

9. Set the dial indicator on a ring gear tooth.   Task Completed ☐

10. Look up the specifications for backlash and record them here.

    _____

11. Rock the ring gear back and forth against the teeth of the pinion gear.   Task Completed ☐

12. Observe the total movement of the indicator. This is the total backlash. Your readings were: _____

13. Measure backlash at four different spots on the ring gear.   Task Completed ☐

14. Describe what needs to happen to correct the backlash.

    _____

    _____

    _____

15. Using knock-in shims or adjusting nuts (depending on axle design), move the ring gear in reference to the pinion gear to achieve proper backlash.   Task Completed ☐

16. When proper backlash is reached, torque the retaining caps to specifications.   Task Completed ☐

17. Recheck the backlash to make sure it is still within specifications.   Task Completed ☐

Problems Encountered

_____

_____

_____

Instructor's Comments

_____

_____

_____

MANUAL DRIVE TRAIN AND AXLES JOB SHEET 31

Differential Case Service

Name _____ Station _____ Date _____

NATEF Correlation

This Job Sheet addresses the following **MAST** task:

> **E.1.13.** Reassemble and reinstall differential case assembly; measure runout; determine necessary action.

Objective

Upon completion of this job sheet, you will be able to reassemble and reinstall a differential case assembly and measure runout.

Tools and Materials

Dial indicator   Gear/bearing puller

Differential holding tool   Hydraulic press

Brass drift

Protective Clothing

Goggles or safety glasses with side shields

Describe the vehicle being worked on:

Year _____ Make _____ Model _____

VIN _____ Engine type and size _____

PROCEDURE

1. With the differential unit still in the housing, set a dial indicator so that its plunger rides on the differential case.   Task Completed ☐

2. By hand, rotate the case and observe the readings on the dial indicator. What did you see during one complete revolution of the case?

   _____

   _____

3. Based on the preceding steps, what are your recommendations?

   _____

   _____

4. If the differential case runout was okay, put the case in a holding fixture.   Task Completed ☐

5. Remove and discard the ring gear bolts.   Task Completed ☐

6. With a brass drift, tap the ring gear loose from the differential case.   Task Completed ☐

7. Rotate the side gears until the pinion gears appear at the case window.   Task Completed ☐

8. Remove the pinion gears, side gears, and thrust washers.   Task Completed ☐

9. Using the appropriate puller, remove the side bearings.   Task Completed ☐

10. To begin reassembly, install the thrust washers on the differential side gears.   Task Completed ☐

11. Position the side gears in the differential case.   Task Completed ☐

12. Place the thrust washers on the pinion gears, and then mesh the pinion gears with the side gears.   Task Completed ☐

13. Install the differential pinion shaft, and then align and install the lock pin into the pinion shaft.   Task Completed ☐

14. Heat the ring gear and install the ring gear onto the differential case.   Task Completed ☐

15. Install and tighten the new ring gear bolts.   Task Completed ☐

16. Use the correct driver and press to install the differential case side bearings.   Task Completed ☐

17. Install the differential with its bearing cones and proper shims.   Task Completed ☐

18. Install the bearing caps and tighten to specifications. What are the specifications? _____

19. Check the preload and backlash of the gear set.   Task Completed ☐

20. Install the axles, c-locks, differential pinion shaft lock pin, and the rear cover.   Task Completed ☐

21. Install the brake drums (or rotors), wheels, and driveshaft.   Task Completed ☐

22. Refill the axle housing with the appropriate type and amount of fluid. What kind of fluid did you put in? How much of it did you put in?

_____

_____

Problems Encountered

_____

_____

_____

Instructor's Comments

_____

_____

_____

MANUAL DRIVE TRAIN AND AXLES JOB SHEET 32

Limited-Slip Differential Diagnostics

Name _____ Station _____ Date _____

NATEF Correlation

This Job Sheet addresses the following **MAST** tasks:

E.2.1. Diagnose noise, slippage, and chatter concerns; determine necessary action.

E.2.2. Measure rotating torque; determine necessary action.

Objective

Upon completion of this job sheet, you will be able to diagnose noise, slippage, and chatter concerns in a limited-slip differential assembly. You will also be able to measure the rotating torque of the assembly.

Tools and Materials

Axle shaft puller

Torque wrench

Protective Clothing

Goggles or safety glasses with side shields

Describe the vehicle being worked on:

Year _____ Make _____ Model _____

VIN _____ Engine type and size _____

PROCEDURE

1. Take the vehicle for a road test. Pay attention to the noise level and action of the differential unit as you make a right and left turn. Describe your findings.

   _____

   _____

2. Now drive the vehicle on a slick surface and accelerate hard enough to lose traction. Pay attention to the noise level and action of the differential unit. Describe your findings.

   _____

   _____

3. Based on the preceding tests, is the unusual noise or action caused by the limited slip function of the differential or is it caused by something else? Explain your answer.

_____

_____

4. A limited-slip differential can be checked for proper operation without removing the differential from the axle housing. Be sure the transmission is in neutral, one rear wheel is on the floor, and the other rear wheel is raised off the floor. Specifications should give the breakaway torque reading required to start the rotation of the wheel that is raised off the floor. What are the specifications?

_____

_____

5. Using a torque wrench, measure the torque required to turn the wheel. What was your measurement and how does it compare with the specification?

_____

_____

Problems Encountered

_____

_____

_____

Instructor's Comments

_____

_____

_____

MANUAL DRIVE TRAIN AND AXLES JOB SHEET 33

Axle Shaft and Bearing Service

Name _____ Station _____ Date _____

NATEF Correlation

This Job Sheet addresses the following **MLR** task:

E.1.1. Inspect and replace drive axle shaft wheel studs.

This Job Sheet addresses the following **AST** tasks:

E.2.1. Inspect and replace drive axle shaft wheel studs.

E.2.2. Remove and replace drive axle shafts.

E.2.3. Inspect and replace drive axle shaft seals, bearings, and retainers.

E.2.4. Measure drive axle flange runout and shaft end play; determine necessary action.

This Job Sheet addresses the following **MAST** tasks:

E.3.1. Inspect and replace drive axle shaft wheel studs.

E.3.2. Remove and replace drive axle shafts.

E.3.3. Inspect and replace drive axle shaft seals, bearings, and retainers.

E.3.4. Measure drive axle flange runout and shaft end play; determine necessary action.

E.3.5. Diagnose drive axle shafts, bearings, and seals for noise, vibration, and fluid leakage concerns; determine necessary action.

Objective

Upon completion of this job sheet, you will be able to diagnose, inspect, remove, and replace axle shafts, bearings, seals, and retainers. You will also be able to inspect and replace drive axle wheel studs and measure drive axle flange runout and shaft end play.

Tools and Materials

Vee blocks	Axle puller
Dial indicator	Slide hammer
Cold chisel	Bearing adapters

Protective Clothing

Goggles or safety glasses with side shields

Describe the vehicle being worked on:

Year _____ Make _____ Model _____

VIN _____ Engine type and size _____

PROCEDURE

1. The diagnosis of any problem should begin with a detailed conversation with the customer to gather as much information about the problem as possible. Ask the customer to carefully describe the problem. This description should include any noises or vibrations, when the problem is evident, and when the problem first became noticeable. If the customer's complaint is based on an abnormal noise or vibration, find out where it is felt or heard. You should also find out if the problem resulted from an event or mishap, such as running into a curb or pothole. Ask the customer about the service history of the car. Often, drive axle problems can be caused by mistakes made by technicians while servicing the vehicle. What are the customer's concerns?

   _____

   _____

   _____

2. If it is possible, take the customer along with you on the road test. Attempt to duplicate the customer's complaint by operating the vehicle under conditions like those during which the problem occurs. Pay careful attention to the vehicle during those and all operating conditions. Keep notes of the vehicle's behavior while driving under various conditions. Note the engine and vehicle speeds at which the problem is most evident. What were the results of the road test?

   _____

   _____

   _____

3. While diagnosing a drive axle noise, it is important to operate the vehicle in all possible conditions and note any changes in noise during each mode. It is also helpful to note how a change in speed affects the noise or vibration. Before assuming that the drive axle is the source of a noise or vibration, make certain the tires or exhaust are not the cause. Inspect the exhaust, suspension, and tires for possible causes of the noise and summarize your findings.

   _____

   _____

4. Poor lubrication in the axle can also be a source of noise, so the lubricant's level and condition should be checked.

   Task Completed ☐

5. If the level is low, the axle should be filled with the proper type and amount of lubricant and the vehicle road-tested again. If the lubricant is contaminated, it should be drained and the axle refilled. Low or contaminated lubricant in a drive axle indicates a leak in the assembly. Therefore, the assembly should be carefully inspected to locate the source of the leak. What did you find?

   _____

   _____

   _____

Checking the Axle Flange

1. Axle shaft replacement is required if there is excessive runout of the axle's flange. To check the runout of the flange, position a dial indicator against the outer flange surface of the axle.   Task Completed ☐

2. Apply slight pressure to the center of the axle to remove the end play in the axle, then zero the indicator.   Task Completed ☐

3. Slowly rotate the axle one complete revolution and observe the readings on the indicator. The total amount of indicator movement is the total amount of axle flange lateral runout. Compare the measured runout with specifications and summarize your findings.

   _____

   _____

4. Inspect the wheel studs in the axle flanges. If they are broken or bent, they should be replaced. What did you find?

   _____

   _____

5. Also check the condition of the threads. If they have minor distortions, run a die over the stud. If the threads are severely damaged, the stud should be replaced. Studs are normally pressed in and out of the flange. Make sure you install the correct size stud. What did you find?

   _____

   _____

Axle Shaft Service

1. If the axle shaft is held in by a retainer, it is not necessary to remove the differential cover to remove the axle. To remove the axle, remove the four bolts that hold the retainer to the backing plate, and then pull the axle out. Normally, the axle shaft will slide out without the aid of a puller, but sometimes a puller is required. Did you face any difficulties?

   _____

   _____

2. A roller bearing supported axle shaft is secured by a C-lock located inside the differential, and the differential cover must be removed to gain access to it. To remove this type of axle, first remove the wheel, brake drum, and differential cover. Then, remove the differential pinion shaft retaining bolt and differential pinion shaft. Now push the axle shaft in and remove the C-lock. The axle can now be pulled out of the housing. Did you face any difficulties?

   _____

   _____

3. When an axle is removed from a housing, the bearing's bore in the axle housing should be inspected and cleaned. If the bore is corroded, sand it lightly with fine emery cloth, and then coat the bore with non–water-soluble grease. This grease will allow the bearing to move slightly in the housing, thereby increasing its service life. The grease will also help prevent corrosion of the bearing's outer race. What did you find?

_____

_____

4. Place the shaft in a pair of Vee-blocks and position a dial indicator at the center of the shaft. Rotating the shaft and observing the indicator will identify any warping or bends in the shaft. Slight indications on the dial indicator may be caused by casting imperfections and do not indicate the need to replace the shaft. What did you find?

_____

_____

5. Ball-type bearings are lubricated with grease packed in the bearing at the factory. This type of bearing is pressed on and off the axle shaft. The retainer ring is made of soft metal and is pressed onto the shaft against the wheel bearing. The ring can be slid off the shaft after it has been drilled into or notched with a cold chisel to break the seal. Did you face any difficulties?

_____

_____

6. Roller axle bearings are lubricated by the gear oil in the axle housing. These bearings are typically pressed into the axle housing. To remove them, the axle must first be removed and then the bearing pulled out of the housing. With the axle out, inspect the area where it rides on the bearing for pits or scores. If pits or score marks are present, replace the axle. What did you find?

_____

_____

7. Tapered-roller axle bearings must be pressed on and off the axle shaft using a press. After the bearing is pressed onto the shaft, it must be packed with wheel bearing grease. After packing the bearing, install the axle in the housing and check the shaft's end play. If the end play is not within specifications, change the size of the bearing shim. Did you face any difficulties?

_____

_____

8. All axle bearings should be inspected for wear or other damage. If the bearings show any evidence of damage, they should be replaced. What did you find?

_____

_____

9. The installation of new axle shaft seals is recommended whenever the axle shafts have been removed. Some axle seals are identified as being either right or left side. When installing new seals, make sure to install the correct seal in each side. How are your seals marked?

_____

_____

10. Coat the outer edge of the new seal with gasket sealer and apply a coat of lubricant to the seal's inner lip. Make sure the lip seal is facing the inside of the axle housing, and then use the correct size seal driver to install the seal squarely in the axle tube. Did you face any difficulties?

_____

_____

11. In most cases, the installation of axle shafts is a simple procedure. Prior to installing the shafts, make sure you installed all of the bearings, seals, and retaining plates on the shaft.

Task Completed ☐

Axle End Play

1. Some axles with tapered-roller bearings require an end play adjustment. To check the end play, position the dial indicator so it can measure the end-to-end movement of the axle.

Task Completed ☐

2. Push the axle into the housing and set the indicator to zero.

Task Completed ☐

3. Pull on the axle and note the indicator's reading. This is the amount of end play. Compare the reading against specifications and correct the end play as necessary. What were the results?

_____

_____

4. End-play adjustments are made by adding or subtracting shims or by turning the adjusting nut. This adjustment is done on one side of the housing but sets the end play for both sides. What did you need to do to correct the end play?

_____

_____

_____

Problems Encountered

_____

_____

_____

Instructor's Comments

_____

_____

_____

MANUAL DRIVE TRAIN AND AXLES JOB SHEET 34

Servicing Transfer Case Shift Controls

Name _____ Station _____ Date _____

NATEF Correlation

This Job Sheet addresses the following **AST/MAST** task:

F.1. Inspect, adjust, and repair shifting controls (mechanical, electrical, and vacuum), bushings, mounts, levers, and brackets.

Objective

Upon completion of this job sheet, you will be able to inspect, adjust, and repair the shifting controls (including mechanical, electrical, and vacuum), as well as the bushings, mounts, levers, and brackets for a transfer case.

Tools and Materials

DMM   Vacuum gauge

Hand-operated vacuum pump   Service information

Protective Clothing

Goggles or safety glasses with side shields

Describe the vehicle being worked on:

Year _____ Make _____ Model _____

VIN _____ Engine type and size _____

PROCEDURE

1. Raise the vehicle on a lift. Is the linkage visible?

   _____

   _____

2. Locate and check the mechanical linkages at the transfer case for looseness and damage. Record your findings here.

   _____

   _____

3. Carefully check the linkage bushings and brackets. Record your findings here.

   _____

   _____

4. Check the mounts for the transfer case and record your findings here.

_____

_____

5. If the mechanical linkage is fine or if there is no direct linkage, proceed to check the electrical controls. Before doing that, is there any evidence that the linkage may be striking the chassis of the vehicle? If so, what can be done to correct that problem?

_____

_____

6. Locate the shift controls and circuit on the wiring diagram. What main components are part of the circuit?

_____

_____

7. Check the control switch for high resistance, which could be the cause of poor shifting. How did you test the switch and what did you find?

_____

_____

8. With the ignition on, use the switch to move from 2WD to 4WD. You should have heard the click of the solenoid or motor. If the clicking was heard, what is indicated?

_____

_____

9. If there was no click, check the circuit fuse. If the fuse is bad, there must be a short in the circuit. What was the condition of the fuse?

_____

_____

10. Raise the vehicle and check for power at the solenoids. Record your findings here.

_____

_____

11. What would be indicated by having the solenoids not working after voltage was applied to them?

_____

_____

12. Visually inspect all electrical connections and terminals. Record your results.

_____

_____

13. If the electrical circuit is fine, check engine manifold vacuum. Record your results and summarize what the results indicate.

_____

_____

14. Visually inspect the vacuum lines for tears, looseness, and other damage. Record your findings here.

_____

_____

15. With the vacuum pump, create a vacuum at the servo and observe its activity and ability to hold a vacuum. Record your results here.

_____

_____

16. If the system has a vacuum motor, apply a vacuum to it and observe its activity and ability to hold a vacuum. Record your results here.

_____

_____

17. If the transfer case has a module that has the capability to display trouble codes, connect a scan tool to the DLC and see if there are any trouble codes stored. List the diagnostic codes along with their meanings here.

_____

_____

18. If trouble codes are stored, briefly describe the diagnostic procedure that will be followed to diagnose the problem.

_____

_____

19. Based on all these checks, what are your service recommendations for this vehicle?

_____

_____

_____

Problems Encountered

_____

_____

_____

Instructor's Comments

_____

_____

_____

MANUAL DRIVE TRAIN AND AXLES JOB SHEET 35

Inspecting Front-Wheel Bearings and Locking Hubs

Name _____ Station _____ Date _____

NATEF Correlation

This Job Sheet addresses the following **MLR** task:

>   **F.1.** Inspect front-wheel bearings and locking hubs; perform necessary action.

This Job Sheet addresses the following **AST/MAST** task:

>   **F.2.** Inspect front-wheel bearings and locking hubs; perform necessary action.

Objective

Upon completion of this job sheet, you will be able to inspect front-wheel bearings and locking hubs on a 4WD vehicle.

Tools and Materials

Lift

Service information

Protective Clothing

Goggles or safety glasses with side shields

Describe the vehicle being worked on:

Year _____ Make _____ Model _____

VIN _____ Engine type and size _____

PROCEDURE

1. Raise the vehicle to a height that allows you to work comfortably around the vehicle's wheels.   Task Completed ☐

2. If the vehicle has locking hubs, rotate a front wheel slightly and move the hub selector to the "lock" position. Did the hub engage with a click?

   _____

3. Does it feel like the axle is now rotating with the wheel?

   _____

4. Do the same thing to the wheel and hub on the other side of the vehicle.   Task Completed ☐

5. Based on these checks, what are your conclusions about the condition of the hubs?

   _____

   _____

6. Grasp the front and rear of a front tire and push it in and out. Do you feel any end play?

_____

_____

7. Do the same to the other front tire. Did you feel any end play?

_____

_____

8. Based on these two checks, do you think the wheel bearings need to be adjusted? Why or why not?

_____

_____

9. If adjustment is needed, remove the tire and wheel assembly.   Task Completed ☐

10. Remove the snap ring at the end of the axle's spindle.   Task Completed ☐

11. Remove the axle shaft spacers, thrust washers, and outer wheel bearing locknut. Keep the spacers and washers in the order they were so you can assemble the unit correctly.   Task Completed ☐

12. Loosen the inner bearing locknut and then fully tighten it to seat the bearings.   Task Completed ☐

13. Spin the brake rotor on its spindle.   Task Completed ☐

14. Loosen the inner bearing locknut approximately one quarter-turn or the amount specified in the service manual.   Task Completed ☐

15. Install the axle shaft spacers and washers, and then install the outer bearing locknut.   Task Completed ☐

16. Tighten the outer locknut to the specified torque. The torque spec is

_____.

17. Install a new snap ring into the end of the axle's spindle.   Task Completed ☐

18. Repeat the wheel bearing adjustment procedure on the other front wheel.   Task Completed ☐

Problems Encountered

_____

_____

_____

Instructor's Comments

_____

_____

_____

MANUAL DRIVE TRAIN AND AXLES JOB SHEET 36

Check Fluid in a Transfer Case

Name _____ Station _____ Date _____

NATEF Correlation

This Job Sheet addresses the following **MLR** task:

F.2. Check drive assembly seals and vents; check lube level.

This Job Sheet addresses the following **AST/MAST** task:

F.3. Check drive assembly seals and vents; check lube level.

Objective

Upon completion of this job sheet, you will be able to inspect a transfer case for leaks and properly check its fluid level.

Tools and Materials

Service information

Protective Clothing

Goggles or safety glasses with side shields

Describe the vehicle being worked on:

Year _____ Make _____ Model _____

VIN _____ Engine type and size _____

Describe the type of system and the model of the transfer case:

_____

PROCEDURE

1. Raise and support the vehicle.   Task Completed ☐

2. Carefully inspect the area around the transfer case for signs of oil leakage. Summarize your findings.

_____

_____

_____

3. Check the connections for the driveshafts to the transfer case. The presence of oil may indicate bad seals. Summarize what you found.

_____

_____

_____

4. Check the mounting point for the transfer case at the transmission, looking for signs of fluid that may indicate a gasket problem at this connection. Summarize what you found.

_____

_____

5. Locate the fill plug on the transfer case. (Refer to the service manual for the location of the plug.)   Task Completed ☐

6. Remove the filler plug.   Task Completed ☐

7. Using your little finger, feel in the hole to determine if you can touch the fluid.   Task Completed ☐

8. If you cannot touch the fluid, refer to the service manual for fluid type. Fill the transfer case. The recommended fluid is _____.

9. If the transfer case is low on fluid, visually inspect it to locate the leaks.   Task Completed ☐

10. If you can reach the fluid with your finger, check the smell, color, and texture of the fluid.   Task Completed ☐

11. If the fluid is contaminated, determine what is contaminating it. Then, correct that problem and drain the fluid from the transfer case and refill it with clean fluid.   Task Completed ☐

Problems Encountered

_____

_____

_____

Instructor's Comments

_____

_____

_____

MANUAL DRIVE TRAIN AND AXLES JOB SHEET 37

Tire Size Changes

Name _____ Station _____ Date _____

NATEF Correlation

This Job Sheet addresses the following **AST/MAST** task:

F.4. Identify concerns related to variations in tire circumference and/or final drive ratios.

Objective

Upon completion of this job sheet, you will be able to understand the effects a change of tire and/or wheel size has on the operation of the vehicle. You will also be able to calculate the difference that a change in size may cause.

Tools and Materials

Service information

Protective Clothing

Goggles or safety glasses with side shields

Describe the vehicle being worked on:

Year _____ Make _____ Model _____

VIN _____ Engine type and size _____

Describe the condition of the tires:

_____

PROCEDURE

1. Carefully check the size of tires on the vehicle. Are they the same size? If not, what and where is the difference?

   _____

   _____

2. When is the only time it is recommended that just one tire be replaced on a vehicle? Where is that tire normally placed on the vehicle?

   _____

   _____

3. When deciding if to replace just two tires, what factors need to be considered and where should those tires be placed on the vehicle?

   _____

   _____

4. When changing the size of the tires to improve handling or fuel economy, what are the main things to consider when determining an acceptable tire size?

_____

_____

5. Most tire width changes affect the overall diameter of the tire. On passenger vehicles, a 3% or less change in tire diameter is acceptable. What can be affected by a greater change than that?

_____

_____

6. Calculate the section height of the following tire sizes by multiplying the tire's aspect ratio by its sectional width. Then, calculate the overall diameter of the tires by multiplying the sectional height by two (this is called the combined sectional height because there are two), then add the diameter of the wheel.

TIRE SIZE	SECTION WIDTH	RIM WIDTH	SECTION HEIGHT	OVERALL DIAMETER
P215/70R14	8.7 inches	6.5 inches	_____	_____
P215/70R15	8.7 inches	6.5 inches	_____	_____
P265/50R14	10.9 inches	8.5 inches	_____	_____
P275/50R15	11.2 inches	8.5 inches	_____	_____
P265/50R16	10.9 inches	8.5 inches	_____	_____

7. The circumference of a tire is the actual length around the tire. It is the distance the tire will travel as it completes one full revolution. Calculate the circumference of these tires by using the formula: pi (π) times the diameter [use a value of 3.14 for π].

TIRE SIZE	OVERALL DIAMETER	CIRCUMFERENCE
P215/70R14	_____	_____
P215/70R15	_____	_____
P265/50R14	_____	_____
P275/50R15	_____	_____
P265/50R16	_____	_____

8. Tread wear has an effect on the circumference of a tire. If a P275/50R15 tire has a tread depth of 11/32-inch when it is new, how much will the circumference decrease when the tire's tread wears down to 3/32-inch?

_____

_____

9. Since the circumference of a tire determines the distance it will travel in one revolution, the number of times the tire will rotate in one mile can be calculated. How will you calculate that?

_____

_____

10. Calculate the revolutions per mile for the following tires: Approximate revolutions per mile for a P215/70R15 tire =

_____

Approximate revolutions per mile for a P275/50R15 tire =

_____

11. Based on the preceding calculations, what would happen to the readings at wheel speed sensors if the P215/70R15 tires were replaced with P275/50R15 tires?

_____

_____

12. What affect would the difference in revolutions per mile have on the vehicle's overall final drive ratio?

_____

_____

13. If a new P275/50R15 tire is placed on the same axle as one that has its tread worn down to 3/32-inch, what possible effects would this have on the handling of the vehicle?

_____

_____

Problems Encountered

_____

_____

_____

Instructor's Comments

_____

_____

_____

MANUAL DRIVE TRAIN AND AXLES JOB SHEET 38

Road-Test a Transfer Case

Name _____ Station _____ Date _____

NATEF Correlation

This Job Sheet addresses the following **MAST** task:

F.5. Diagnose noise, vibration, and unusual steering concerns; determine necessary action.

Objective

Upon completion of this job sheet, you will be able to road-test a vehicle and determine the operating condition of a transfer case.

Tools and Materials

Vehicle with four-wheel drive

Describe the vehicle being worked on:

Year _____ Make _____ Model _____

VIN _____ Engine type and size _____

Describe the transfer case controls and list the type of transfer case found on the vehicle:

_____

PROCEDURE

Road-test the vehicle and attempt to operate the vehicle in all of the modes of the transfer case.

1. Does the transfer case make noise in:

   a. Low drive in rear-wheel drive?   yes _____ no _____ n/a _____

   b. High drive in rear-wheel drive?   yes _____ no _____ n/a _____

   c. Low drive in four-wheel drive?   yes _____ no _____ n/a _____

   d. High drive in four-wheel drive?   yes _____ no _____ n/a _____

2. Does it make noise as you turn corners with it in four-wheel drive?
   (Note: On dry pavement tires may slip when turning and the vehicle may be hard to steer. Do not operate for an extended period of time on dry pavement in four wheel drive)
   yes _____ no _____

3. Does it make noise as you drive down the road in high gear?
   yes _____ no _____

4. Does it make a noise when the transfer case is in neutral?
   yes _____ no _____

Problems Encountered

_____

_____

_____

Instructor's Comments

_____

_____

_____

MANUAL DRIVE TRAIN AND AXLES JOB SHEET 39

Servicing 4WD Electrical Systems

Name _____ Station _____ Date _____

NATEF Correlation

This Job Sheet addresses the following **MAST** task:

F.6. Diagnose, test, adjust, and replace electrical/electronic components of four-wheel-drive systems.

Objective

Upon completion of this job sheet, you will be able to diagnose, test, adjust, and replace the electrical and electronic components of four-wheel-drive systems.

Tools and Materials

DMM   Small mirror

Flashlight   Air nozzle

Scan tool   Service information

Protective Clothing

Goggles or safety glasses with side shields

Describe the vehicle being worked on:

Year _____ Make _____ Model _____

VIN _____ Engine type and size _____

PROCEDURE

1. Describe the complaints for the transfer case in question and list them.

   _____

   _____

2. Check all electrical connections to the transfer case. Make sure they are tight and not damaged. Record your findings.

   _____

   _____

3. Release the locking tabs of the connectors and disconnect them, one at a time, from the transfer case. Carefully examine them for signs of corrosion, distortion, moisture, and transmission fluid. A connector or wiring

harness may deteriorate if automatic transmission fluid reaches it. Using a small mirror and flashlight may help you get a good look at the inside of the connectors. Record your findings.

_____

_____

4. Inspect the entire wiring harness for tears and other damage. Record your findings.

_____

_____

5. Because the operation of the engine, transmission, and transfer case are integrated through the control computer, a faulty engine sensor or connector may affect the operation of all of these. With a scan tool, retrieve any diagnostic trouble codes (DTCs) saved in the computer's memory. Were there any present? If so, what is indicated by them?

_____

_____

6. The engine control sensors that are the most likely to cause shifting problems are the throttle-position sensor, manifold absolute position (MAP) sensor, and vehicle speed sensor. Locate these sensors and describe their location.

_____

_____

_____

7. Remove the electrical connector from the throttle-position (TP) sensor and inspect both ends for signs of corrosion and damage. Record your findings.

_____

_____

8. Inspect the wiring harness to the TP sensor for evidence of damage. Record your findings.

_____

_____

9. Check both ends of the three-pronged connector and wiring at the MAP sensor for corrosion and damage. Record your findings.

_____

_____

10. Check the condition of the vacuum hose. Record your findings.

_____

_____

11. Check the speed sensor's connections and wiring for signs of damage and corrosion. Record your findings.

   _____

   _____

12. Record your conclusions from the visual inspection.

   _____

   _____

   _____

   _____

13. Remove the selector switch and check its action with an ohmmeter. Move the switch to all possible positions and observe the ohmmeter. Compare the action of the switch to the circuit shown in the wiring diagram. Summarize the condition of the switch.

   _____

   _____

   _____

14. Locate the transfer case shift motor or solenoid.   Task Completed ☐

15. Carefully inspect this for damage and poor mounting. Record your findings here.

   _____

   _____

16. Check the solenoid or motor following the procedures given by the manufacturer. Summarize this check and your results.

   _____

   _____

   _____

17. If the solenoid or motor must be removed for testing, make sure it is mounted correctly and the mounting bolts tighten to specs when you reinstall it.   Task Completed ☐

18. Check the service manual and identify all switches located on the transfer case. List these here.

   _____

   _____

   _____

19. Determine the function of each switch. Is it a grounding switch or does it complete or open a power circuit?

   _____

   _____

   _____

20. An ohmmeter can be used to identify the type of switch being used and to test the operation of the switch. There should be continuity when the switch is closed and no continuity when the switch is open. Record your findings.

   _____

   _____

21. Apply air pressure to the part of the switch that would normally be exposed to oil pressure and check for leaks. Record your findings.

   _____

   _____

22. The transfer case may be equipped with a speed sensor. Typically, the speed sensor is a permanent magnetic (PM) generator. Locate the speed sensor and describe its location.

   _____

   _____

23. With the vehicle raised on a lift, allow the wheels to be suspended so they are free to rotate.   Task Completed ☐

24. Set your DMM to measure AC voltage.   Task Completed ☐

25. Connect the meter to the speed sensor.   Task Completed ☐

26. Start the engine and put the transmission in gear. Slowly increase the engine's speed until the vehicle is at approximately 20 mph, and then measure the voltage at the speed sensor. Record your findings.

   _____

   _____

27. Slowly increase the engine's speed and observe the voltmeter. The voltage should increase smoothly and precisely with an increase in speed. Record your findings.

   _____

   _____

28. A speed sensor can also be tested with it out of the vehicle. Connect an ohmmeter across the sensor's terminals. What is the measured resistance?

   _____

   _____

29. Locate the specifications for the sensor and compare your readings with specifications.

_____

_____

Problems Encountered

_____

_____

_____

Instructor's Comments

_____

_____

_____

MANUAL DRIVE TRAIN AND AXLES JOB SHEET 40

Disassemble and Reassemble a Transfer Case

Name _____ Station _____ Date _____

NATEF Correlation

This Job Sheet addresses the following **MAST** task:

> **F.7.** Disassemble, service, and reassemble transfer case and components.

Objective

Upon completion of this job sheet, you will be able to disassemble, service, and reassemble a transfer case.

Tools and Materials

Pry bar
Torque wrench
Service Information

Protective Clothing

Goggles or safety glasses with side shields

Describe the vehicle being worked on:

Year _____ Make _____ Model _____

VIN _____ Engine type and size _____

Transfer case type _____

> **NOTE:** *This procedure is a typical procedure. Follow the service information on your particular transfer case for the exact procedure necessary.*

PROCEDURE

1. If not previously drained, remove the drain plug and allow the oil to drain, then reinstall the plug.   Task Completed ☐

2. Loosen the flange nuts.   Task Completed ☐

3. Remove the two output shaft yoke nuts, washers, rubber seals, and output yokes from the case.   Task Completed ☐

4. Remove the four-wheel-drive indicator switch from the cover.   Task Completed ☐

5. Remove the wires from the electronic shift harness connector.   Task Completed ☐

6. Remove the speed sensor retaining bracket screw, bracket, and sensor.   Task Completed ☐

7. Remove the bolts securing the electric shift motor and remove the motor. Note the location of the triangular shaft in the case and the triangular slot in the electric motor.   Task Completed ☐

8. Loosen and remove the front case to rear case retaining bolts.   Task Completed ☐

9. Separate the two halves by prying between the pry bosses on the case.   Task Completed ☐

10. Remove the shift rail for the electric motor.   Task Completed ☐

11. Pull the clutch coil off the main shaft.   Task Completed ☐

12. Pull the 2WD/4WD shift fork and lock-up assembly off the main shaft.   Task Completed ☐

13. Remove the chain, driven sprocket, and drive sprocket as a unit.   Task Completed ☐

14. Remove the main shaft with the oil pump assembly.   Task Completed ☐

15. Slip the high-low range shift fork out of the inside track of the shift cam.   Task Completed ☐

16. Remove the high-low shift collar from the shift fork.   Task Completed ☐

17. Unbolt and remove the planetary gear mounting plate from the case.   Task Completed ☐

18. Pull the planetary gear set out of the mounting plate.   Task Completed ☐

19. Install the input shaft and front output shaft bearings into the case.   Task Completed ☐

20. Apply a thin bead of sealer around the ring gear housing.   Task Completed ☐

21. Install the input shaft with planetary gear set and tighten the retaining bolts to specifications. What are the specifications?

    _____

    _____

22. Install the high-low shift collar into the shift fork.   Task Completed ☐

23. Install the high-low shift assembly into the case.   Task Completed ☐

24. Install the main shaft with oil pump assembly into the case.   Task Completed ☐

25. Install the drive, driven sprockets, and chain into position in the case.   Task Completed ☐

26. Install the shift rails.   Task Completed ☐

27. Install the 2WD/4WD shift fork and lock-up assembly onto the main shaft.   Task Completed ☐

28. Install the clutch coil onto the main shaft.   Task Completed ☐

29. Clean the mating surface of the case.   Task Completed ☐

30. Position the shafts and tighten the case halves together. Tighten attaching bolts to specifications. What are the specifications?

    _____

    _____

31. Apply a thin bead of sealer to the mating surface of the electric shift motor.   Task Completed ☐

32. Align the triangular shaft with the motor's triangular slot.   Task Completed ☐

33. Install the motor over the shaft. Wiggle the motor to ensure it is fully seated on the shaft.

Task Completed ☐

34. Tighten the motor's retaining bolts to specifications. What are the specifications?

_____

_____

35. Reinstall the wires into the connector and connect all electric sensors.

Task Completed ☐

36. Install the companion flanges' seal, washer, and nut. Then, tighten the nut to specifications. What are the specifications?

_____

_____

37. Summarize how successful you were.

_____

_____

_____

_____

Problems Encountered

_____

_____

_____

Instructor's Comments

_____

_____

_____

— NOTES —

— **NOTES** —

— NOTES —

— NOTES —

— **NOTES** —